新视角职业英语
立体化规划教材

NEW PERSPECTIVE VOCATIONAL ENGLISH

计算机专业英语

赵中颖 / 主　编
宋丹　董爽 / 副主编

ENGLISH FOR COMPUTER ENGINEERING

人民邮电出版社
北京

图书在版编目（CIP）数据

计算机专业英语 / 赵中颖主编. -- 北京：人民邮电出版社，2016.2（2019.6重印）
新视角职业英语立体化规划教材
ISBN 978-7-115-41595-0

Ⅰ．①计… Ⅱ．①赵… Ⅲ．①电子计算机－英语－职业教育－教材 Ⅳ．①H31

中国版本图书馆CIP数据核字(2016)第004990号

内 容 提 要

本书将"就业导向"和"工学结合"作为编写的指导理念，章节内容设计包括了从基础理论到具体应用多个方位的专业知识。全书共 10 个单元，涵盖了计算机硬件、软件以及网络 3 大方面内容。全书内容前后呼应，融会贯通。

本书适合职业院校、成人高校的计算机科学与技术、网络工程、信息管理与信息系统、软件工程等专业学生用作教材，对于 IT 从业人员也有较强的参考价值。

♦ 主　　编　　赵中颖
　　副主编　　宋　丹　董　爽
　　责任编辑　　刘　琦
　　责任印制　　张佳莹　杨林杰
♦ 人民邮电出版社出版发行　　北京市丰台区成寿寺路 11 号
　　邮编　100164　电子邮件　315@ptpress.com.cn
　　网址　http://www.ptpress.com.cn
　　北京隆昌伟业印刷有限公司印刷
♦ 开本：787×1092　1/16
　　印张：14　　　　　　　　　　　　　2016 年 2 月第 1 版
　　字数：346 千字　　　　　　　　　　2019 年 6 月北京第 7 次印刷

定价：42.00 元

读者服务热线：(010) 81055256　印装质量热线：(010) 81055316
反盗版热线：(010) 81055315
广告经营许可证：京东工商广登字20170147号

Preface

　　高等职业教育担负着培养高素质技能型人才和专门人才的重要任务。本系列教材从高等技术应用型人才培养的总体目标出发，结合学生毕业后的工作实际，力求向学生提供其未来工作岗位所需要的通用英语和专业英语的知识与技能，提高学生实际应用英语的能力。

　　一、编写宗旨

　　1. 深入贯彻《高职高专教育英语课程教学基本要求（试行）》和《高等职业教育英语课程教学要求（试行）》，以培养学生实际应用英语的能力为目标，侧重职场环境下语言交际能力的培养，训练学生在生产、管理、服务第一线所需的语言交际能力和应对各种涉外局面的语言应用能力，旨在培养高素质的技能型人才。

　　2. 充分考虑高职高专学生的英语基础现状，最大限度地使教材内容与学生的基础相适应，与培养目标相一致。

　　3. 在深入细致调研的基础上，将"就业导向"和"工学结合"作为教材编写的指导理念，由长期从事一线教学，拥有丰富经验的教师对本书内容进行精心设计、反复论证。

　　二、编写特色

　　1. 本书打破传统，将通用英语与专业英语有机结合，颠覆了原有专业英语只强调阅读理解和翻译能力培养的教学理念，以交际为特色，突出听说能力的训练，并以此促进学生阅读和翻译能力的提高。

　　2. 本书总体结构和单元结构层次分明、重点突出，体现"实用为主、够用为度"及循序渐进的原则，具有较强的系统性和连贯性，有助于激发学生作为教学主体参与课堂活动的积极性。

　　3. 练习设计有的放矢，重难、难点突出，通过多种形式反复操练，使听、说、读、写、译5项基本技能在同一主题下通过多种练习形式反复循环、巩固和加强，使学生真正做到"学一点、练一点、会一点、用一点"，把课内与课外有机结合起来，把教与学有机结合起来。

　　4. 本书设计新颖、图文并茂，配有大量与主题相关且启发性强的图片，为语言学习提供了形象立体的训练情景，激发了学生学习和使用语言的兴趣。

　　5. 本书打破传统的学科教材模式，采用直观呈现内容为主的形式生动地突出行业特点，不强调专业知识的系统性和完整性，注重岗位工作情境设置，培养学生在真实职场环境下实际应用语言的能力。

　　三、内容简介

　　本书的内容包含了从基础理论到具体应用的各类专业知识，共10个单元，涵盖了计算机硬件、软件以及网络3大方面。

　　1. 导入部分

　　倡导听说领先，通过浏览与单元主题相关的图片以及听简短的对话了解该单元的内容；以讨论的方式进入学习，调动学生的主观能动性。

　　2. 对话

　　通过职场环境下的真实语境，调整小组成员并反复操练情景对话；通过记忆训练和模

Preface

仿练习，有效地促成行业语言版块的形成。

3. 课文

文章语言真实规范，题材新颖，密切联系实际，没有艰深晦涩、专业难度过大的内容。课文篇幅短小实用，一般在200字左右，避免长篇课文带来的沉重感，从而消除学生的畏惧感。

4. 翻译技巧

介绍常见的科技英语翻译方法与技巧，通过理论知识讲解和大量实例展示，提高学生的翻译能力。

5. 写作

将常见的应用文体与商务公文写作相结合，通过一系列的范例学习和写作实践来培养学生阅读和模拟套写常用应用文的能力。

6. 语法

通过适当的讲解和大量的实践练习，查缺补漏，巩固和提高学生的英语语法知识和语言应用能力。

7. 文化点滴

拓展与主题相关的背景知识，增加学生对英语的学习兴趣，开阔视野。

四、任务分工

本书由辽宁机电职业技术学院的赵中颖、宋丹、董爽编写。其中，第6单元~第9单元（包括参考答案）及前言由赵中颖编写，第3单元~第5单元（包括参考答案）由宋丹编写，第1单元、第2单元、第10单元（包括参考答案）由董爽编写。

本书在编写过程中得到了各方面的支持和帮助，并参考了大量文献资料。在此，我们向所有的作者表示感谢。同时，也感谢领导、同事及家人的默默支持和不断鼓励！

由于编者水平有限，尽管在编写过程中已竭尽全力，多次审校，但书中难免有疏漏和不妥之处，恳请广大读者批评指正。

编　者

2015年12月

CONTENTS

Unit One Computer History 1
    Section One Warming Up 1
    Section Two Real World 2
    Section Three Brighten Your Eyes 7
    Section Four Translation Skills 11
    Section Five Writing 15
    Section Six Grammar 17

Unit Two Hardware 23
    Section One Warming Up 23
    Section Two Real World 24
    Section Three Brighten Your Eyes 29
    Section Four Translation Skills 34
    Section Five Writing 37
    Section Six Grammar 40

Unit Three Office Software 45
    Section One Warming Up 45
    Section Two Real World 45
    Section Three Brighten Your Eyes 50
    Section Four Translation Skills 54
    Section Five Writing 58
    Section Six Grammar 60

Unit Four Computer Network 65
    Section One Warming Up 65
    Section Two Real World 66
    Section Three Brighten Your Eyes 71
    Section Four Translation Skills 75
    Section Five Writing 79
    Section six Grammar 81

CONTENTS

Unit Five Network Security 87
    Section One Warming Up 87
    Section Two Real World 88
    Section Three Brighten Your Eyes 92
    Section Four Translation Skills 97
    Section Five Writing 100
    Section Six Grammar 102

Unit Six Virus 107
    Section One Warming Up 107
    Section Two Real World 107
    Section Three Brighten Your Eyes 112
    Section Four Translation Skills 117
    Section Five Writing 120
    Section Six   Grammar 123

Unit Seven Website 129
    Section One Warming Up 129
    Section Two Real World 130
    Section Three Brighten Your Eyes 135
    Section Four Translation Skills 139
    Section Five Writing 144
    Section Six Grammar 146

Unit Eight Search Engine 151
    Section One Warming Up 151
    Section Two Real World 152
    Section Three Brighten Your Eyes 156
    Section Four Translation Skills 161
    Section Five Writing 164
    Section Six Grammar 166

CONTENTS

Unit Nine Network and System Administrator 171
    Section One Warming Up 171
    Section Two Real World 172
    Section Three Brighten Your Eyes 177
    Section Four Translation Skills 182
    Section Five Writing 186
    Section Six Grammar 188

Unit Ten IT Jobs 193
    Section One Warming Up 193
    Section Two Real World 195
    Section Three Brighten Your Eyes 200
    Section Four Translation Skills 204
    Section Five Writing 209
    Section Six Grammar 211

参考文献 216

Unit One

Computer History

Section One Warming Up

Pair work

Can you name anyone who have made great difference for computers? What did they do? Talk with your partner and discuss what the people in the following pictures did? Match the names with the corresponding pictures.

Bill Gates
Steve Paul Jobs
Douglas C. Engelbart
John Vincent Atanasoff

1. _____

2. _____

3._____

4._____

Section Two Real World

Apple Inc. now is an icon in the electronic products world and the company employs an apple with one bite missing as its **logo**. Do you know why the founder of Apple—Steve Jobs designed its logo like that?

Listen and act the following dialogue.

Apple Inc. Logo

Alex: Hi, Blair. Did you see the movie that I've introduced to you called *The **Imitation** Game*?

Blair: Oh, yes, of course. It is actually a **biography** film about Alan Turing whom I have never heard before, to be honest.

Alex: Well, it is true that few people know about him. But as a solid Apple fan, he is a character we cannot ignore.

Blair: Why is that?

Alex: Jobs **admired** Alan Turing, who was believed to make a great achievement for the development of computer science.

Blair: I get that from the movie. He invented one machine to **decode** the Enigma for the Allies during World War Ⅱ ,which made the computer-like machine called the Turing machine.

Alex: That's quite right. But due to his **sexual preference**, he was **arrested** for treatment. He couldn't **endure** the process and **committed suicide** by eating a poisoned apple.

Blair: That's really a **tragic** story.

Alex: Yes. In memory of this great man, Jobs used an apple with one bite missing as his company logo.

Blair: You're truly an Apple fan!

Words & Expressions

logo [ˈləʊgəʊ]	n.	a company emblem or symbol 商标
imitation [ˌɪmɪˈteɪʃ(ə)n]	n.	a copy that is represented as the original 模仿
biography [baɪˈɒgrəfɪ]	n.	an account of the series of events making up a person's life 传记；档案
solid [ˈsɒlɪd]	adj.	reliable; having a strong basis 可靠的；结实的
ignore [ɪgˈnɔː]	vt.	refuse to acknowledge 忽视；不理睬
admire [ədˈmaɪə]	vt.	feel admiration for 钦佩
decode [diːˈkəʊd]	vt.	convert code into ordinary language [计]解码
sexual [ˈsekʃʊəl]	adj.	of or relating to or characterized by sexuality 性别的
preference [ˈpref(ə)r(ə)ns]	n.	a strong liking 偏爱，倾向
arrest [əˈrest]	vt.	the act of apprehending 逮捕
endure [ɪnˈdjʊə]	vt.	put up with something or somebody unpleasant 忍耐
commit [kəˈmɪt]	vt.	perform an act, usually with a negative connotation 犯罪
suicide [ˈs(j)uːɪsaɪd]	n.	the act of killing oneself 自杀
tragic [ˈtrædʒɪk]	adj.	very sad 悲剧的
a biography film		传记电影
make a great achievement		取得巨大成就
in memory of...		纪念……

Proper Names

The Imitation Game	《模仿游戏》
the Turing machine	图灵机

Find Information

Task 1 Read the dialogue carefully and then answer the following questions.

1. What's the name of the movie introduced to Blair?

2. What's the type of the movie and what is it about?

3. What does the logo of Apple Inc. look like?
_____

4. Did Turing contribute to Word War Ⅱ?
_____

5. How did Alan Turing die?
_____

Task II Read the dialogue carefully and then decide whether the following statements are true (T) or false (F).

(　) 1. Turing helped the Allies to decode the Enigma machine during the First World War.
(　) 2. Jobs admired Turing, therefore he designed the company logo with an apple missing one bite.
(　) 3. Due to the arrestment, he killed himself.
(　) 4. Blair is a solid Apple fan.
(　) 5. The story of Turing is tragic.

Words Building

Task I Choose the best answer from the four choices A, B, C and D.

(　) 1. These apples _____ in size from small to medium.
　　A. yield　　　　　B. vary　　　　　C. exist　　　　　D. apply

(　) 2. She's _____ some brilliant schemes to double her income.
　　A. come up with　　　　　B. gone over for
　　C. come out for　　　　　D. gone on with

(　) 3. You've got to _____ in yourself, or you'll never succeed.
　　A. assure　　　　　B. confirm　　　　　C. believe　　　　　D. imagine

(　) 4. I have never seen _____ film before.
　　A. such a moving　　　　　B. such moving
　　C. so moving　　　　　D. such a moved

(　) 5. Had it not been for his help, I _____ the task within so little time.
　　A. can't complete　　　　　B. couldn't complete
　　C. wouldn't complete　　　　　D. couldn't have complete

Task II Fill in each blank with the proper form of the word given.

1. People _____ diamonds with crystal. (imitation)
2. That man has done a rather _____ job of imitating the human brain. (admire)
3. Would you _____ milk or coffee? (preference)
4. To do this, we must have _____ at the highest level of government. (commit)
5. It is _____ pottering away valuable time. (suicide)

Unit One Computer History

Task III Match the following English terms with the equivalent Chinese.

A — advanced applications
B — 2-D code
C — Bluetooth
D — wireless mouse
E — connection device
F — digital notebook
G — expansion card
H — history files
I — internet telephony
J — operation system

1. ( ) 无线鼠标
2. ( ) 数字笔记本
3. ( ) 二维码
4. ( ) 扩展卡
5. ( ) 高级应用
6. ( ) 历史文件
7. ( ) 操作系统
8. ( ) 连接设备
9. ( ) 蓝牙
10. ( ) 网络电话

Cheer Up Your Ears

Task I Listen and write down what you've heard. Then read and recite till you can use them fluently.

1. Did you see the movie that I've _____ to you called *The Imitation Game*?
2. It is actually a _____ film about Alan Turing whom I have never heard before, to be _____ .
3. But as a solid Apple fan, he is a _____ we cannot ignore.
4. Jobs _____ Alan Turing, who was believed to make a great _____ for the development of computer science.
5. I get that from the movie. He invented one machine to _____ the Enigma for the Allies during World War Ⅱ.
6. He couldn't _____ the process and committed _____ by eating a poisoned apple.
7. That's really a _____ story.
8. In _____ of this great man, Jobs used an apple with one bite missing as his company logo.

Task II Listen and fill in the blanks with the missing words and role-play the conversation with your partner.

Buying a Computer

A: Can I help you?
B: Yes, I want to buy a ___1___.
A: What do you want to do with it?
B: Oh, I'm not sure. I don't know much about computers.
A: Do you want to play computer games or write ___2___?
B: What's the ___3___?
A: Well, for game playing, it's better if your computer has a bigger ___4___ and a better ___5___ video card.

B: I guess Pentium IV would be good.

A: Yes, but if you're ___6___ about game playing effects, I suggest you buy a bigger ___7___.

B: Yeah, that's important. Then how about the big one over there?

A: Oh, you are so smart! That one is 20 inches and is made in Japan. But the monitor with a big screen will ___8___ more space.

B: I don't care. I've got a very large living room.

A: That's all right. Here you are. The total ___9___ would be $ 850.00. Are you sure you're going to buy it now?

B: Oh, sorry, not right now. Next week, maybe. Thanks for your help.

A: My ___10___, madam. See you next time.

B: See you!

Task III Listen to 5 recorded questions and choose the best answer.

1. A. Yes, there is nothing.   B. My pleasure.
   C. I'd like a cup of coffee.   D. You're welcome.
2. A. I prefer having a rest.   B. I see.
   C. I'm not sure.   D. I love it.
3. A. Yes, I like.   B. No, thanks.
   C. I'm not sure.   D. How about you?
4. A. She's all right.   B. She feels bad.
   C. She can go to school.   D. Twice per week.
5. A. He was ill.   B. He liked music.
   C. He knew it.   D. He did it yesterday.

Table Talk

Task I Complete the following dialogue by translating Chinese into English orally.

Talk about Memory Capacity

A: Where have you been?

B: I have been to the computer component store. What's up?

A: I want to upgrade my computer by changing the old ___1___ (硬盘).

B: Oh dear. It's not easy. Is there something wrong with the old one?

A: Not exactly. But it often causes me trouble, especially when I run some large programs. I think its ___2___ (内存)is too small.

B: How much ___3___ (随机储存器) do you have?

A: 80M.

B: My goodness! Your computer must be ___4___ (过时的).

A: So I think it's time for me to upgrade it. What's your suggestion?

B: Do you plan to write papers, ___5___ (记录) your business on a spreadsheet , surf the

Unit One Computer History

Internet or just play computer games?
A: I plan to do all of those.
B: Then, you need at least 40G. It may cost you a lot of money.
A: That's all right. Thanks for your advice.
B: My pleasure.

Task II **Pair-work. Role-play a conversation about the purchase of a computer with your partner.**

Situation
Suppose you are going to buy a computer. Ask your friend for some advice such as hardware, software, etc. Now talk to your partner.

Section Three Brighten Your Eyes

Pre-reading questions:
1. When and by whom was the first computer invented?
2. Can you give a brief introduction of computer history?

Development of Computers

Nothing **epitomizes** modern life better than the computer. For better or worse, computers have **infiltrated** every **aspect** of our society. Today computers do much more than simply computers. To fully understand and **appreciate** the impact computers have on our lives and **promises** they hold for the future, it is important to understand their **evolution**.

1. First-generation Computers (1946 ~ 1959)

First-generation computers were characterized by the fact that operating instructions were made-to-order for the specific task for which the computer was to be used. Each computer had a different binary-coded program called a machine **language** that told it how to operate.

2. Second-generation Computers (1959 ~ 1964)

It was the stored program and programming language that gave computers the **flexibility** to finally be cost **effective** and **productive** for business use. New types of careers (programmer, analysts and computer systems expert) and the **entire** software industry began with the second—generation computers.

3. Third-generation Computers (1964 ~ 1971)

Computers became even smaller as more components were **squeezed** onto the integrated circuit (IC) chip. Another third-

generation development included the use of an operating system that allowed machines to run many different programs at once with a central program.

4. Fourth- generation Computers (1971-Present)

After the integrated circuits, the only place to go down was in size, that is, large scale integration (LSI) could fit hundreds of components onto one chip.

5. Fifth-generation computers (Present and Future)

Defining the fifth-generation computers is somewhat difficult because this field is in its **infancy**. Computers today have some attributes to the fifth.

Words & Expressions

epitomize [ɪ'pɪtəmaɪz]	vt.	embody the essential characteristics of or be a typical example of 概括
infiltrate ['ɪnfɪltreɪt]	vt.	cause to enter by penetrating the interstices 使渗入
aspect ['æspekt]	n.	a distinct feature or element in a problem 方面
appreciate [ə'priːʃɪeɪt]	vt.	recognize with gratitude 欣赏；感激
impact ['ɪmpækt]	n.	a forceful consequence; a strong effect 影响
promise ['prɒmɪs]	n.	grounds for feeling hopeful about the future 许诺
evolution [ˌiːvə'luːʃ(ə)n]	n.	a process in which something passes by degrees to a different stage 演变
language ['læŋgwɪdʒ]	n.	communication by word of mouth 语言
flexibility [ˌfleksɪ'bɪlɪtɪ]	n.	the property of being flexible; easily bent or shaped 灵活性
effective [ɪ'fektɪv]	adj.	producing or capable of producing an intended result or having a striking effect 有效的
productive [prə'dʌktɪv]	adj.	producing or capable of producing goods or crops, esp in large quantities （能）生产的
entire [ɪn'taɪə]	adj.	constituting the full quantity or extent 全部的
squeeze [skwiːz]	vt.	press firmly 挤；紧握
define [dɪ'faɪn]	vt.	give a definition for the meaning of a word 定义
infancy ['ɪnf(ə)nsɪ]	n.	the early stage of growth or development 初期
for better or worse		不论好坏；不管怎样
made-to-order		订制的
Large scale integration(LSI)		大规模集成

Find Information

Task I Read the passage carefully and then answer the following questions.

1. Do computers today simply compute?

2. What is the binary-code that the 1st generation computer has called?
_____

3. Which industry began with the 2nd generation computer?
_____

4. Where is the fouth-generation computers go down?
_____

5. Why is it difficult to define the fifth-generation computers?
_____

Task II Read the passage carefully and then decide whether the following statements are true (T) or false (F).

(   ) 1. First-generation computers were characterized by the fact that operating instructions were made-to-demand for the specific task.
(   ) 2. The stored program and programming language gave the second generation computers the flexibility to finally be cost effective and productive for business use.
(   ) 3. Third-generation computers are the smallest of all the computers.
(   ) 4. Large scale integration makes computers even smaller in size.
(   ) 5. Computers today have no attributes to the fifth-generation computers.

Words Building

Task I Translate the following phrases into English.

1. 二进制编码程序_____
2. 机器语言_____
3. 存储程序_____
4. 集成电路芯片_____
5. 大规模集成_____
6. 在起步阶段_____

Task II Translate the following sentences into Chinese or English.

1. To fully understand and appreciate the impact computers have on our lives and promises they hold for the future, it is important to understand their evolution.
_____

2. Each computer had a different binary-coded program called a machine language that told it how to operate.
_____

3. Another third-generation development included the use of an operating system that allowed machines to run many different programs at once with a central program.
_____

4. After the integrated circuits, the only place to go down was in size, that is, large scale integration (LSI) could fit hundreds of components onto one chip.
_____

5. 没有东西能够像计算机一样集中体现现代生活。
_____

6. 不论怎样，计算机已渗透到我们社会生活的每一方面。
_____

7. 随着更多的部件被压缩在了集成芯片上，计算机变得更小。
_____

8. 要定义第五代计算机在某种程度上来说很难，因为这一领域仍在起步阶段。
_____

Task III Choose the best answer from the four choices A, B, C and D.

( ) 1. He is thought to _____ many popular books.
  A. write   B. have written   C. have writing   D. have been written

( ) 2. Tom came to the dinner without _____ .
  A. inviting   B. to invite   C. invite   D. being invited

( ) 3. Hearing the news, the boy burst into _____ .
  A. laugh   B. laughing   C. laughter   D. laughable

( ) 4. Violence on TV may turn out to have an _____ on some young people.
  A. effect   B. efficiency   C. influence   D. impression

( ) 5. He hardly ever leaves the house after ten at night, _____ ?
  A. doesn't he   B. hasn't he   C does he   D. has he

Task IV Fill in the blanks with the proper form of the word given.

1. The knowledge of anatomy adds to the _____ of works of art. (appreciate)
2. We're trying to bring along several _____ young football player. (promise)
3. It has passed through an interesting procedure of _____ . (evolutionary)
4. We have to resign ourselves to fate since we cannot think out an _____ remedy. (effect)
5. The fire was _____ caused by their neglect of duty. (entire)

Cheer up Your Ears

Task I Listen and write down what you've heard. Then read and recite till you can use them fluently.

1. For better or worse, computers have _____ every aspect of our society.
2. To fully understand and _____ the impact computers have on our lives and promises they hold for the future, it is important to understand their _____ .
3. First-generation computers were _____ by the fact that operating instructions were made-to-

order for the specific task for which the computer was to be used.

4. Each computer had a different binary-coded program called a machine _____ that told it how to operate.
5. It was the stored program and programming language that gave computers the _____ to finally be cost effective and productive for business use.
6. New types of _____ (programmer, analysts and computer systems expert) and the entire software _____ began with the second-generation computers.
7. Computers became even smaller as more components were _____ onto the integrated circuit (IC) chip.
8. Defining the fifth-generation of computers is somewhat difficult because the field is in its _____ .

Task II Listen to 5 short dialogues and choose the best answer.

1. A. He didn't like that computer.
   B. He didn't find what he liked.
   C. The price of the computer was too high.
   D. That type of computer was sold out.
2. A. Go to a lecture.   B. Go to a concert.
   C. Go shopping.   D. Go sightseeing.
3. A. Write a letter for the woman. B. Take the woman to the office.
   C. Drive the woman home.   D. Finish the report for the woman.
4. A. She was tired reading it.   B. She liked it very much.
   C. She didn't think much of it. D. She wasn't interested in it.
5. A. When he can receive the order. B. What the order number is.
   C. When he should send the order. D. What's wrong with the order.

Table Talk

Group Work

Talk about the development of the computer with your partner and try to describe the first computer you have.

_____

_____

Section Four Translation Skills

科技英语的翻译方法与技巧——英汉语言的不同

翻译是将一种语言文字的思想内容和风格用另一种语言的习惯表达方式确切而完整地重新表达出来的过程。专业英语的翻译就是要将专业英语文献转换成汉语译文，使人们能够借助汉语译文准确无误地了解英语专业著作所阐述的科学理论和工程技术内容。为了做好翻译，首先应对语言的差别有所认识。其次，翻译的过程就是理解和表达的过程，主要包括对原文含义的正确理解和用切的汉语句子准确表达原文的含义。

专业英语的翻译标准应是准确而流畅的。科技文献的翻译，准确是第一位的，在准确的基础上进一步要求流畅。要特别注意：逻辑和术语正确，结构严谨，表达简练。科技文献主要是论述事理，其逻辑性强，结构严谨，术语繁多，所以译文必须满足概念清晰、逻辑正确、数据无误、文字简练、通顺易懂的要求，尤其是术语、定义、定理、公式、算式、图表、结论等更要注意准确恰当。从事专业英语翻译的人应该注意提高自己的外语水平、汉语水平和专业知识水平。

一、句子结构的差别

汉语是分析性语言，依靠词序表达各成分之间的关系，词序是严格的，而英语的词序比较灵活。英语和汉语的句子中主语、谓语、宾语和表语的词序大体一致，但是定语和状语的位置有时相同，有时不同，在翻译时应该注意汉语的习惯。英语中可作定语的有单词、短语和定语从句，以单词作定语一般前置，以短语和从句作定语一般后置。汉语的定语一般在被修饰词之前。

1. 主句和从句的次序

英、汉两种语言在主句和从句的次序以及几个从句之间的次序上存在着差别，主要表现在各个句子的时间顺序和逻辑顺序上。

（1）时间顺序

汉语中，一般按照主句和从句所叙述的事情发生的先后排列，而英语表示时间的从句在复合句中的位置比较灵活。

例如：

The primary dendrites solidify before eutectic reaction when temperature decreases.

当温度降低时，初生枝晶在共晶反应之前凝固。

（2）逻辑顺序

在主句和从句的逻辑顺序方面，英语对主、从句先后次序的要求与汉语相同。例如：

Without a feeder, the pressure of the mould cavity and the inside pressure of the sand mould become negative, the mould will then collapse because it loses the pressure difference necessary for supporting its formation.

如果没有冒口，型腔的压力和砂型的内压将变成负值，这时砂型由于失去了支持其形成所必需的压力差将发生崩溃。

2. 句子结构的差别

在英语句子中，各个词、词组、子句之间需要采用介词、连词等连接，以表达其相互的语法关系，因此英语的句子结构较为严谨。而在汉语的句子中，各个词、词组、子句则是按照时间顺序和逻辑关系贯穿起来的，往往无需采用任何连词连接。因此，汉语的句子结构较松，文句较简洁。例如：

We are interested in producing a component that has the proper shape and properties and permitting the component to perform for its expected lifetime.

我们感兴趣的是制造具有特定形状和性能的零件，且允许这个零件在其预期的寿命内工作。

二、词汇的差别

英语和汉语在词的意义、词的搭配、词序等词汇现象上也存在着差别。

三、词的搭配上的不同

英语和汉语在词的搭配上存在着较大的差别，主要表现在形容词和名词、动词和名词、动词和副词等搭配上。

如 heavy 一词作为形容词，它的基本词义是"沉重的、重型的、重的、大量的"等，但当与下面不同的名词搭配时，按照汉语的搭配习惯，则应有不同的译法，如：

heavy alloy	高密度合金	heavy casting	大型铸件
heavy machine	重型机械	heavy iron	镀锌薄板
heavy mill	大型轧钢机	heavy plant	重工业厂
heavy scale	厚氧化皮	heavy oil	重油
heavy current	强电流	heavy load	重载
heavy traffic	拥挤的交通	heavy repair	大修
heavy fire	猛烈的炮火	heavy crops	丰收
heavy storm	大风暴	heavy snow	大雪
heavy wire	粗导线		

又如动词 burn 的词义原为"燃烧、灼热、发光、消耗"等，而在机械工程专业中，当这个英语动词与下述不同介词或名词搭配时，按照汉语的习惯，应分别译为：

burn away	烧光	burn-on	黏砂
burn back	熔接	burned sand	焦砂
burn in	腐蚀	burned-on-sand	包砂
burn-off	蒸发	burning-off	机械黏砂

在英语句子中，状语若是一个单词，其位置根据所修饰的词语的不同而不同。若单词状语所修饰的是形容词，该单词状语一般前置；若单词状语所修饰的是动词，一般后置；若用来修饰某个状语的单词状语是一个表示程度的状语，则既可能前置，也可能后置。若状语是一个短语而且用来修饰动词时，该状语既可放在动词之前，也可放在动词之后。但汉语句子中的状语一般都放在它所修饰的词语前面。

词义方面的不同。

不论在英语中还是在汉语中，绝大多数的词都是多义词。在词义上英汉两种语言的差别主要表现在词义的对应上，有以下 3 种不同的情况：

（1）英语的词义与汉语的词义对等。例如：

steel	钢	aluminum	铝
solidification	凝固	microstructure	显微组织
temperature	温度		

（2）一个英语词所表达的意义与几个具有不同的词义甚至词性的汉语词分别对应。例如，metal 一词在汉语中与之对应的有名词"金属，合金，五金，成分，轴承，碎石料"等，还有动词"用金属包镀，覆以金属，用碎石铺路面"等。又如 cast 一词在汉语中有"铸造，浇注，浇灌，熔炼，投射，筹划，预测，分类整理"等名词，以及具有相同含义的动词与之对应。

（3）英语的词义与汉语的词义部分对应。英汉两种词义在所概括的范围上有大小之分。如英语中的 furnace 一词，在汉语中可泛指各种类型的"炉子"，包括鼓风炉、热处理炉、干锅炉，甚至家用燃气炉、煤炉等，而 cupola 仅指化铁用的"冲天炉"。反过来，汉语中的"特性"一词,在英语中有 characteristic、property、performance、behavior 等几个具有类似意义,但用于不同场合的词与之对应。

Listening

Task I **Listen to 2 conversations and choose the best answer.**

Conversation 1

1. A. Three weeks ago. B. Last month.
   C. Last week. D. Last Monday.
2. A. Install more machines. B. Test the machines.
   C. Buy more machines. D. Sell the machines.

Conversation 2

3. A. TV sets. B. Music players.
   C. Desk top computers. D. Digital cameras.
4. A. 270 dollars. B. 280 dollars.
   C. 290 dollars. D. 300 dollars.
5. A. 2 percent. B. 3 percent.
   C. 4 percent. D. 5 percent.

Task II **Listen to the passage and fill in the blanks with the missing words or phrases.**

Modern technology has a big influence on our daily life. New devices are widely used today. For example, we have to __1__ the Internet every day. It is becoming more and more __2__ to nearly everybody. Now it's time to think about how the Internet influences us, what __3__ it has on our social behaviour. The Internet has __4__ changed our life; there is no doubt about that. I think that the Internet has changed our life in a __5__ way.

Extensive Reading

Directions: After reading the passage, you are required to complete the outline below briefly.

Alan Turing

Alan Mathison Turing (23 June, 1912 ~ 7 June, 1954) was a British pioneering computer scientist, mathematician, logician, cryptanalyst, philosopher, mathematical biologist, and marathon and ultra distance runner. He was highly influential in the development of computer science, providing a formalisation of the concepts of "algorithm" and "computation" with the Turing machine, which can be considered a model of a general purpose computer. Turing is widely considered to be the father of theoretical computer science and artificial intelligence.

During the Second World War, Turing worked for the Government Code and Cypher School (GC&CS) at Bletchley Park, Britain's codebreaking centre. For a time he led Hut 8, the section is responsible for German naval cryptanalysis. He devised a number of techniques for breaking German ciphers, including the improvements to the prewar polish bomb method. Turing's pivotal role in cracking intercepted coded messages enabled the Allies to defeat the Nazis in many crucial engagements, including the Battle of the Atlantic; it has been estimated that the work at Bletchley Park shortened the war in Europe by as many as two to four years.

Turing was prosecuted in 1952 for homosexual acts, when such behavior was still criminalised in the UK. He accepted treatment with oestrogen injections (chemical castration) as an alternative to prison. Turing died in 1954, 16 days before his 42nd birthday, from cyanide poisoning. An inquest determined his death a suicide, but it has been noted that the known evidence is equally consistent with accidental poisoning. In 2009, following an Internet campaign, British Prime Minister Gordon Brown made an official public apology on behalf of the British government for "the appalling way he was treated". Queen Elizabeth II granted him a posthumous pardon in 2013.

Alan Turing's Profile
Gender: ___1___
Born in ___2___
Died in ___3___ from ___4___
Nationality: ___5___
Occupations: ___6___, mathematician, logician, cryptanalyst, philosopher, mathematical biologist, and ___7___ and ultra distance runner.
Achievements: a. Father of ___8___ and ___9___
b. shortened the war in ___10___ by as many as two to four years.

Section Five Writing

Business Card 名片

名片是在社交时用于简单介绍个人情况的小卡片，上面通常印有个人的姓名、头衔、职务、单位以及电话、住址、传真、电子邮件等联系方式。名片是现代社会中最为广泛应用的一种交流工具。

Read and understand the sample business card.

Sample

The Great Wall Suzhou Electronics Co., Ltd.	
Wang Ming	
Software Dept. Manager / Electronics Engineer	
Address: No.11 Changjiang Road, Suzhou	Tel: 0512-82873128
Post Code: 215001	Fax: 0512-82873292
E-mail: W-ming@yahoo.com.cn	Mobile: 15906132397

范例

长城苏州电子有限公司
王明
软件部经理 / 电子工程师
地址：苏州市长江路 11 号　　电话：0512-82873128
邮编：215001　　　　　　　　传真：0512-82873292
邮箱：W-ming@yahoo.com.cn　手机：15906132397

名片的书写规则

1. 名片上的文字应工整、对称。有的名片上的文字向名片的左边靠齐，有的位于名片的中间。通常，通信地址放在左下角，电话号码放在右下角。

2. 由于名片大小的限制，在表达方式上尽量用缩略语。英国的公司 Company 经常简写为 Co., 有限公司为 Co., LTD. 美国的公司常用 Corp. 或者 Inc. 的缩写表示，全拼分别为 Corporation 和 Incorporation。Dept. 表示部门 Department，Rd. 表示路 Road，St. 表示 Street，Sq. 表示广场 Square，Add. 表示地址 Address，Tel. 表示电话 Telephone。P.C. 表示邮编 Post code。

常用语

职务职称

professor 教授；associate professor 副教授；president 大学校长；dean 系主任；principal 中学校长；school master 小学校长；director 导演，主任，处长；senior lecturer 高级讲师；chief engineer 总工程师；senior engineer 高级工程师；senior accountant 高级会计师；designer 设计师；senior economist 高级经济师；technician 技师，技术员；official 公务员；chairman of the board 董事长；chief executive officer (CEO) 执行总裁；reporter 记者；general manager 总经理；sales manager 销售经理；marketing manager 营销经理；personnel manager/director 人事主管；secretary-general 秘书长；general editor 总编辑

单位名称

bureau 局；department 系，部；section 处，科；institute 所；office 室；agency 社；group company 集团公司；broadcasting station 电台；TV station 电视台

Practice

Read the following information and choose any information necessary for the layout of a business card. Then design a business card. 阅读下面的信息，选择必要的内容，设计一张名片。

Name: William White
Gender: male
Date of Birth: Feb.23, 1968
Marital status: married
Address: 12 Nicholson Avenue, Canberra City, Australia

Post Code: ACT2601
Company Name: Sanderson Motor Group
Title: General Manager
Telephone: 632574721
Fax No.: 632574723
E-mail address: William@sanderson.com

Section Six Grammar

动词不定式 Infinitive

一、动词不定式的基本形式

	主动语态	被动语态
一般式	to do	to be done
完成式	to have done	to have been done
进行式	to be doing	/

其否定形式是在 to 前加上 not。

二、动词不定式的基本用法

（一）作主语

1. 动词不定式作主语时，句子的谓语动词常用单数。例如：

To lean out of the window is dangerous. 身子探出窗外很危险。

2. 不定式作主语时，如果不定式过长，为避免头重脚轻的现象，常用形式主语 it 放前，而真正的主语置于句末。例如：

To learn a foreign language is necessary for us.

It's necessary for us to learn a foreign language.

学一门外语对我们来说很必要。

3. 常用形式主语 it 的句型有：

It's impossible/important/easy/hard/necessary for sb. to do sth.

It's wise/nice/kind/clever/wrong/foolish/careless/polite of sb. to do sth.

当句中作表语的形容词修饰整个不定式短语时用 for sb. to do sth.

当句中作表语的形容词表示逻辑主语的特性时用 of sb. to do sth. 例如：

It's hard for him to go to sleep. 他很难入睡。

It's foolish of her to do that. 她那样做是愚蠢的。

（二）作宾语

1. 以下动词和短语后，只能跟不定式作宾语。

afford（付得起），agree（同意），aim(力求做到)，appear（显得），arrange（安排），ask（要求），attempt（试图），care（想要），choose（选择），claim（声称），condescend（屈尊），consent（准许），decide（决定），demand（要求），determine（决心），endeavor（竭力），expect（期待），fail（未履行），help（帮助），hesitate（犹豫），hope（希望），learn（学会），manage（设法），neglect（疏忽），offer（主动提出），plan（计划），prepare（准备），pretend（假装），proceed（接着做），promise（答应），prove（证明），refuse（拒

绝），resolve（解决），seem（觉得好像），swear（发誓），tend（往往会），threaten（预示），undertake（承诺），volunteer（自愿做），vow（发誓），want（想要），wish（希望，would like（想要），would love（想要），make up one's mind /be determined（决定）……
例如：

The driver failed to see the other car in time. 司机没能及时看见另一辆车。

I happen to know the answer to your question. 我碰巧知道你那道问题的答案。

2. 有一些要求接双宾语的动词，如 tell, teach, show, advise, find out 等可跟一个带连接代词或副词的不定式做宾语。例如：

He showed me what to do next. 他给我展示了接下来应该怎么做。

Please tell me how to learn English well. 请告诉我怎样才能学好英语。

但要注意 why 不能用 why to do 做宾语。

（三）作宾语补足语

1. 带动词不定式做宾补的动词、词组有 ask, tell, order, advise, want, teach, force, forbid, beg, allow, expect, cause, invite, know, ask for, call on, depend on, wait for 等。例如：

Father will not allow us to play in the street. 父亲不让我们在街上玩耍。

We are waiting for the bus to come. 我们正在等公共汽车的到来。

2. 感官动词 see, watch, look at, observe ,notice, hear, listen to, feel 等以及 3 个使役动词 let, make, have 要用不带 to 的不定式作宾补，但在被动语态中要加上 to。（let, have 一般不用于被动语态）。例如：

Though he had often made his little sister cry, today he was made to cry by his little sister. 尽管他经常把妹妹惹哭，但是今天他被妹妹惹哭了。

注意：help 后的不定式可有可无 to。例如：Can you help me (to) carry the box?

3. think，consider，believe，suppose，find，prove，imagine，understand 等后接宾语再加 to be 再加形容词（adj.）或名词（n.）等作宾补时可省略 to be。例如：

Many people consider him (to be) an honest man.

很多人认为他是一个诚实的人。

（四）作主语补足语

不定式做主语补足语时，和主语构成一种逻辑上的主谓关系。如：

He was not allowed to enter the room for being late.

他因为迟到了被拒之门外。

She is said to be studying abroad. 据说，她正在国外读书。

（五）作表语

表示"愿望、目的"的词作主语时，其表语常用不定式形式。如：

His wish was to become a famous scientist.

他的愿望是成为一名著名科学家。

The purpose of the test is to check the students' English.

测试的目的是检查学生的英语水平。

（六）作状语

1. 表示目的。相当于 in order to do, so as to do 等句型。例如：

I get up early (in order) to catch the first bus. 为了赶头班车我起得很早。

To pay off the debt, she worked day and night.

为了还债，她没日没夜的工作。

作目的状语的不定式肯定式 to/in order to，so as to，其否定形式应用 in order not to 或 so as not to。另外，so as to/ so as not to 不可置于句首。

2. 表示结果。相当于 too ... to do, enough to do 等句型。例如：

The child is too young to go to school. 那个孩子太小没到上学的年龄。

（七）作定语

1. 不定式通常在序数词、形容词最高级、the last 等修饰的名词或代词后作定语。例如：

He loves parties. He is always the first to come and the last to leave.

他喜欢聚会。总是第一个来最后一个走。

Would you like something to eat? 你想要点吃的吗？

2. 不定式为不及物动词时，其后须加上适当的介词，构成及物动词短语。

例如：

I am looking for a room to live in. 我正在找一间屋子住。

I need a piece of paper to write on. 我需要一张写字用的纸。

但是，当不定式作定语且所修饰的名词是 time，place 或 way 时，不定式后的介词一般要省去。

例如： He had no money and no place to live. 他没有钱也没有地方住。

3. 不定式短语作定语时，在某些结构中虽然表示被动，但用的却是主动形式。例如：

I have a question to ask you. 我有个问题要问你。

We still have many difficulties to smooth away.

我们仍然有很多问题要解决。

（八）不定式省略 to 的一些常见情况：

1. why/why not 后

Why not cancel the meeting? 为什么不取消会议？

2. would rather 后

He would rather stay at home by himself. 他宁愿自己待在家里。

3. had better 后

You'd better stop talking. 你最好停止讲话。

4. can not but 后

I can not but tell him the truth. 我不得不告诉他真相。

5. but 前是实义动词 do 的某种形式时，后面的不定式不带 to。例如：He wants to do nothing but go out. 他什么都不想做，只想出去。

6. 作感官动词、使役动词的宾补。

Let them go at once. 让他们马上走。

He made me cry yesterday. 他昨天把我惹哭了。

7. 由 and, or 和 than 连接的几个不定式时，只在第一个不定式前加 to。例如：He forgot to go to her home and give her the important letter.

他忘了去她家并把一封重要的信件给她。

Exercises

I. Choose the best answer to each of the following items.

1. The thief tried to escape _____.
   A. to be caught   B. being caught   C. to catch   D. catching
2. He told the children _____ make so much noise.
   A. don't   B. don't to   C. to not   D. not to
3. When the lecture was going on, he came in _____.
   A. with no one notice   B. without being noticed
   C. with no one to notice   D. without anyone noticing
4. Visitors cannot help _____ by the beautiful natural scenery.
   A. impressing   B. impressed   C. to be impressed D. being impressed
5. I prefer _____ ather than _____.
   A. read; watch   B. to read; watch
   C. reading; to watch   D. to read; to watch
6. _____ won't be of much help.
   A. Monitor going   B. Monitor will go
   C. Monitor's going   D. For monitor to go
7. —Will the Smiths go abroad this summer?—No, they finally decided _____.
   A. to   B. not going   C. not to   D. not to be going
8. Tom as well as his two friends _____.
   A. are worthy of teaching   B. are worth to teaching
   C. are worth teaching   D. is worth teaching
9. We don't know _____ next.
   A. to do what   B. what to do   C. how we do it   D. what we do
10. He is always the first _____ to the office.
    A. comes   B. coming   C. to come   D. who came
11. The little girl could do nothing but _____.
    A. wait   B. to wait   C. waiting   D. to be waiting
12. I felt it an honor _____ to speak here.
    A. to ask   B. asking   C. to be asked   D. having asked
13. Dr. Smith devoted 20 years to _____ this subject.
    A. study   B. to have studied   C. studying   D. to study
14. The famous novel is said _____ into Chinese.
    A. to have translated   B. to be translate
    C. to have been translated   D. to translate
15. The teacher has his students _____ a composition every other week.
    A. to write   B. written   C. writing   D. write

II. Complete the following sentences.
1. He was so angry that he couldn't say anything.

   He was too angry _____ _____ anything.
2. I don't know when we will have the meeting.

   I don't know when _____ _____ the meeting.
3. That you read English in the morning is very important.

   It is very important _____ you _____ _____ English in the morning.
4. He was so strong that he could lift the stone.

   He was strong _____ _____ _____ the stone.
5. They got up early so that they could get there in time.

   They got up early _____ _____ _____ _____ there in time.
6. He hopes that he can visit the Great Wall.

   He hopes _____ _____ the Great Wall.
7. I saw him go into the room.

   He was seen _____ _____ into the room.
8. He stopped and had a look at me.

   He stopped _____ _____ a look at me.
9. The box is so heavy that I can't carry it.

   The box is too heavy _____ me _____ _____.
10. "Lie down!" the boy said to his dog.

    The boy ordered his dog _____ _____ down.

CULTURE TIPS

扫一扫了解冯·诺依曼

Unit Two

Hardware

Section One Warming Up

A complete PC system consists of two basic parts— hardware and software. Hardware is the physical part of the system that can be seen and touched which is under control of the software. We usually divide them into four groups: storage devices, input devices, output devices and networking devices.

Pair work

Can you name some parts of a computer that are referred to as hardware? Match the following parts with the pictures.

CPU	Mouse
Keyboard	Mainboard
Printer	Monitor

1. _____   2. _____

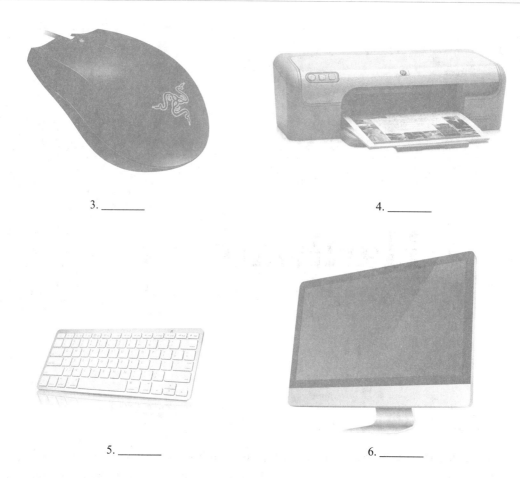

3. _____   4. _____

5. _____   6. _____

Section Two Real World

There are so many **components** that you can seldom see if you are a common user, such as the power supply, motherboard, Central Processing Unit (CPU), RAM and ROM, hard drive, sound card and video card. These are common ones you might have known.

Listen and act the following dialogue which is about the difference of RAM and ROM.

RAM and ROM

Alex: Hi, Blair. It is really difficult for me to tell the difference between RAM and ROM. Can you explain that to me?

Blair: Sure, Alex, it's my pleasure. Well, it is true that beginners may get **confused**. RAM stands for Random **Access Memory**.

Alex: Oh, yes. It records new information, doesn't it?

Blair: Yes, it remembers what you tell them and can **switch** to remember new information, but when the computer is turned off, RAM forgets everything you told it.

Alex: I see. So, that's why it is so **essential** to save the work on a computer. What does ROM **represent** then?

Blair: It stands for Read Only Memory, which is good at remembering but cannot change its mind.

Alex: So I can't change any information there?

Blair: Yes, it's like reading a **library** book. There is lots of information built into it but you can **alter** nothing.

Alex: RAM is more like a **journal** that I can write or **erase** information into but ROM is more like a borrowed book where I can't change anything. I've understood the differences. Thank you so much!

Blair: It's quite all right.

Words & Expressions

component [kəm'pəʊnənt]	n.	an abstract part of something 成分；[电子]元件
confused [kən'fjuːzd]	adj.	unable to think with clarity or to act intelligently 困惑的；糊涂的
random ['rændəm]	adj.	taken haphazardly [数]随机的；任意的
access ['ækses]	n.	the right to enter 进入；使用权
memory ['mem(ə)rɪ]	n.	something that is remembered 记忆；[计]存储器
switch [swɪtʃ]	vt.	make a shift in or exchange of 转换
essential [ɪ'senʃ(ə)l]	adj.	absolutely necessary 基本的；必要的
represent [reprɪ'zent]	vt.	take the place of or be parallel 代表
library ['laɪbrərɪ]	n.	a room where books are kept 图书馆
alter ['ɔːltə]	vt.	cause to change; make different 改变；更改
journal ['dʒɜːn(ə)l]	n.	a daily written record of (usually personal) experiences and observations 日报；日记
erase [ɪ'reɪz]	vt.	remove from memory or existence 抹去；擦除
tell the difference		区别
turn off		关闭
change one's mind		改变某人的想法

Proper Names

RAM　　　　　　　　　　　　随机存储器
ROM　　　　　　　　　　　　只读存储器

Find Information

Task I Read the dialogue carefully and then answer the following questions.

1. If you're a common user of computer, are video card and sound card the most commonly seen parts?
   _____

2. What's Alex's problem?
   _____

3. What does RAM stand for?
   _____

4. What does ROM represent?
   _____

5. Which of the two is like a library book?
   _____

Task II Read the dialogue carefully and then decide whether the following statements are true (T) or false (F).

(　) 1. Blair knows the difference between RAM and ROM.
(　) 2. Read Only Memory records new information.
(　) 3. RAM can't change to remember new information.
(　) 4. ROM can still keep the information put in whether the computer is turned off or not.
(　) 5. It is important to save the documents before switching off the computer.

Words Building

Task I Choose the best answer from the four choices A, B, C and D.

(　) 1. There _____ no time left, we had to change our plan.
   A. being　　　　B. had　　　　C. would be　　　　D. has

(　) 2. I _____ you had returned the book yesterday.
   A. want　　　　B. would rather　　C. liked　　　　D. had better

(　) 3. I invited Tom and Ann to dinner, but _____ of them came.
   A. nether　　　　B. both　　　　C. either　　　　D. none

(　) 4. The government must take some measures _____ the city's pollution.
   A. to control　　B. controls　　C. controlling　　D. to be controlled

(　) 5. My advisor encouraged _____ a summer course to improve my writing skills.
   A. for me taking　B. me taking　C. for me to take　D. me to take

Task II Fill in each blank with the proper form of the word given.

1. It is _____ for other people if you use an old picture for your profile. (confuse)

2. The veteran soldier has fought many _____ battles. (memory)
3. Without further _____ or manipulation, it does not change its nature. (alter)
4. It's really difficult to tell the _____ between butter and mayonnaise. (differ)
5. Please _____ my name from the list. (eraser)

Task III Match the following English terms with the equivalent Chinese.

A — keyboard　　　　　　　1.（　）扫描仪
B — printer　　　　　　　　2.（　）喇叭
C — scanner　　　　　　　　3.（　）打印机
D — case　　　　　　　　　4.（　）液晶显示屏
E — webcam　　　　　　　　5.（　）存储器
F — speaker　　　　　　　　6.（　）网络摄影机
G — memory　　　　　　　　7.（　）机箱
H — LCD　　　　　　　　　8.（　）键盘
I — compact disk　　　　　　9.（　）分辨率
J — resolution　　　　　　　10.（　）光盘

Cheer up Your Ears

Task I Listen and write down what you've heard. Then read and recite till you can use them fluently.

1. There are so many _____ that you can seldom see if you are a common user.
2. It is really difficult for me to tell the _____ between RAM and ROM.
3. It is true that beginners may get _____ .
4. RAM stands for _____ Access Memory.
5. It remembers what you tell them and can _____ to remember new information.
6. So, that's why it is so _____ to save the work on a computer.
7. What does ROM _____ then?
8. There is lots of information built into it but you can _____ nothing.
9. RAM is more like a _____ that I can write or _____ information into but ROM is more like a borrowed book where I can't change anything.

Task II Listen and fill in the blanks with the missing words and role-play the conversation with your partner.

A: Here you go! It comes with a Fujitus __1__ and a 17 inch Panasonic __2__ .
B: Those boxes are so big!
A: Plus a keyboard and a mouse. The __3__ and an Asus __4__ are built-in.
B: Are you sure you haven't forgotten anything?

A: Oh, yeah, __5__ and an HP LaserJet printer.

B: I was just joking.

A: It's great value for the money. But maybe you want to add some __6__.

B: First things first. Is there a warranty?

A: A one-year warranty, free parts and __7__.

B: What about after that?

A: You can still return it to the factory for repair, but for a __8__.

B: All this for just 7,000 yuan. I can't believe it.

A: Plus 1,000 yuan for the __9__.

B: That's good. I'll take it.

A: Then __10__ yuan altogether.

B: Here you are.

A: Thanks a lot.

Task III Listen to 5 recorded questions and choose the best answer.

1. A. Yes, quite difficult.   B. No, I didn't.
   C. Here you are.   D. Yes, you can.
2. A. Don't worry about it.   B. Why not.
   C. Here you are.   D. Of course, please.
3. A. No, you needn't.   B. No, you can't.
   C. I can't believe it.   D. Thank goodness.
4. A. It's very kind of you.   B. It's useful one.
   C. It's 15 dollars.   D. It's a thick book.
5. A. Here it is.   B. Sure it will.
   C. Benz, of course.   D. Yes, I like it.

Table Talk

Task I Complete the following dialogue by translating Chinese into English orally.

Asking for Advice on Computer Care

A: Excuse me, Ms Nance. May I __1__ （问你一个问题）?

B: Of course. Go ahead.

A: Do you know how to keep computers clean?

B: That sounds like a simple topic but computers require special cleaning even on the outside.

A: What do you mean by special cleaning on the outside?

B: It means that you have to keep your system free from __2__ （灰尘、污垢和液体）.

A: What kind of liquids must be kept away?

B: All kinds, especially __3__ （玻璃清洁液）.

A: Why?

B: Because mixing liquids and electronic components can cause serious __4__ (触电).

A: That sounds terrible. And what else should I pay attention to?

B: It's best to stay clear of the computer during storms.

A: Do you mean it's dangerous to use the computer during storms?

B: Right. Remember lightening could __5__ (被传导到) your computer.

A: Thanks for the advice.

B: You're welcome.

Task II Pair-work. Role-play a conversation about the hardware system of a computer with your partner.

Situation

Suppose you're going to buy a computer. Ask the salesman about the hardware system of a computer that you are interested in. Now you are talking to your salesman.

Section Three Brighten Your Eyes

Pre-reading questions:

1. Can you say any main parts of computers? What are they?
2. What's CPU short for?

The CPU

In stand-alone (non-networked) usage situations, computer performance is largely determined by three computer **components**: the CPU, RAM and the computer **monitor**.

Like a brain, the CPU (Central **Processing** Unit) controls information and tells other parts what to do. It processes computer instructions to perform work, just as an **automobile** engine processes fuel to produce power. CPU performance is measured in terms of the **maximum** number of instructions the processor may carry out per second, **expressed** in **megahertz** (MHz). One MHz equals one million instruction cycles per second. Current-generation CPUs are **incredibly** fast. If a CPU operates at or above 350MHz, most PC users do not **appreciate** much of a difference if additional speed is added. Obviously, the difference

is there, but it is **subtle**, not **obvious**.

Most modern CPUs are microprocessors, meaning they are contained on a single **integrated** circuit (IC) chip. An IC that contains a CPU may also contain memory, **peripheral** interfaces, and other components of a computer; such integrated devices are variously called microcontrollers or systems on a chip (SoC). Some computers employ a multi-core processor, which is a single chip containing two or more CPUs called "cores", in that **context**, single chips are sometimes referred to as "sockets". **Array** processors or **vector** processors have **multiple** processors that operate in **parallel**, with no unit considered central.

Words & Expressions

monitor	['mɒnɪtə]	n.	display produced by a device that takes signals and displays them on a television screen or a computer monitor 显示屏
process	['prəʊses]	vt.	particular course of action intended to achieve a result 处理；加工
automobile	['ɔːtəməbiːl]	n.	a motor vehicle with four wheels 汽车
maximum	['mæksɪməm]	n.	the largest possible quantity [数] 极大；最大量
megahertz	['megəhɜːts]	n.	one million periods per second 兆赫
incredibly	[ɪnˈkredəblɪ]	adv.	not easy to believe 难以置信地；非常地
appreciate	[əˈpriːʃɪeɪt]	vt.	recognize with gratitude 欣赏；感激
subtle	['sʌt(ə)l]	adj.	be difficult to detect or grasp the mind 微妙的；精细的
obvious	['ɒbvɪəs]	adj.	easily perceived by the senses or grasped by the mind 明显的
integrated	['ɪntɪgreɪtɪd]	adj.	formed or united into a whole 综合的；完整的
peripheral	[pəˈrɪf(ə)r(ə)l]	adj.	electronic equipment connected by the CPU of a computer 外围的
interface	['ɪntəfeɪs]	n.	a program that controls a display for the user (usually on a computer monitor) and that allows the user to interact with the system 界面
array	[əˈreɪ]	n.	an orderly arrangement 数组；阵列；排列
vector	['vektə]	n.	a variable quantity that can be resolved into components 矢量
multiple	['mʌltɪpl]	adj.	having or involving or consisting of more than one part 多重的
parallel	['pærəlel]	n.	something having the property of being analogous to something else 平行线
be determined by...			由……决定
in terms of...			依据，按照
carry out			执行

Unit Two　Hardware

an integrated circuit (IC) chip	集成电路芯片
be referred to as...	被称为……

Find Information

Task I　**Read the passage carefully and then answer the following questions.**

1. What determines the performance of a computer in the non-networked situation?
   _____

2. What does a CUP do?
   _____

3. What is used to measure the CUP performance?
   _____

4. What does "microprocessor" mean?
   _____

5. Except CPU, what else may an IC contain?
   _____

Task II　**Read the passage carefully and then decide whether the following statements are true (T) or false (F).**

(　) 1. In the passage, a CPU is not only described as a brain, but also as a car engine.
(　) 2. One MHz equals one billion instruction cycles per second.
(　) 3. A multi-core processor means that a single chip containing two or more CPUs.
(　) 4. Single chips are sometimes referred to as "sockets" in a multi-core processor.
(　) 5. Array processors or vector processors have multiple processors that operate in parallel, with certain unit considered central.

Words Building

Task I　**Translate the following phrases into English.**

1. 中央处理器_____
2. 计算机性能_____
3. 集成电路芯片_____
4. 外围接口_____
5. 多核处理器_____

Task II　**Translate the following sentences into Chinese or English.**

1. In stand-alone (non-networked) usage situations, computer performance is largely determined by three computer components.
   _____
   _____

2. CPU performance is measured in terms of the maximum number of instructions the processor may carry out per second, expressed in megahertz (MHz).
_____

3. Some computers employ a multi-core processor, which is a single chip containing two or more CPUs called "cores".
_____

4. Array processors or vector processors have multiple processors that operate in parallel, with no unit considered central.
_____

5. 如同大脑一样，中央处理器控制信息并指示其他部件行动。
_____

6. 它处理计算机指令执行工作，就和汽车的引擎把油料变成动力一样。
_____

7. 在不影响操作的情况下，刀具伸出部分越短越好。
_____

8. 在那种情况下，单一的芯片被叫作"套接口"。
_____

Task III Choose the best answer from the four choices A, B, C and D.

(　　) 1. _____ in the queue for half an hour, Tom suddenly realized that he had left his wallet at home.
　　A. To wait　　B. Waiting　　C. Having waited　　D. To have waited

(　　) 2. The book is worthy of _____, I think.
　　A. reading　　B. be read　　C. being read　　D. read

(　　) 3. I'd like to buy a house — modern, comfortable, and _____ in a quite neighborhood.
　　A. in all　　B. above all　　C. after all　　D. at all

(　　) 4. The reporters hurried to the airport, only _____ that the film star had left.
　　A. to tell　　B. to be told　　C. telling　　D. told

(　　) 5. Having been ill in bed for nearly a month, he had a hard time _____ the exam.
　　A. pass　　B. to pass　　C. passed　　D. passing

Task IV Fill in the blanks with the proper form of the word given.

1. Some computers employ a multi-core _____ . (process)
2. I can't _____ the beauty of Picasso. (appreciation)
3. An _____ chip that contains a CPU may also contain memory, peripheral interfaces, and other components of a computer. (integrate)
4. He is _____ to as the first man to eat the crab in computer history. (refer)
5. Jerry is _____ to make great difference of the house left by his grandpa .(determine)

Cheer up Your Ears

Task I Listen and write down what you've heard. Then read and recite till you can use them fluently.

1. In stand-alone (non-networked) usage situations, computer performance is largely _____ by three computer components.
2. Like a brain, the CPU controls _____ and tells other parts what to do.
3. It processes computer instructions to perform work, just as an automobile engine _____ fuel to produce power.
4. CPU performance is _____ in terms of the maximum number of instructions the processor may carry out per second.
5. One MHz _____ one million instruction cycles per second.
6. Current-generation CPUs are _____ fast.
7. Sure, the difference is there, but it is subtle, not _____.
8. Most modern CPUs are microprocessors, meaning they are contained on a single _____ circuit (IC) chip.
9. In that context, single chips are sometimes _____ to as "sockets".
10. Array processors or vector processors have multiple processors that operate in parallel, with no unit _____ central.

Task II Listen to 5 short dialogues and choose the best answer.

1. A. She didn't know the time.　　B. She forgot her class.
   C. She didn't catch the bus.　　D. The bus was late.
2. A. 9:00.　　B. 9:50.
   C. 8:45.　　D. 8:15.
3. A. In a store.　　B. On a plane.
   C. In the hospital.　　D. At the theater.
4. A. Vegetables.　　B. Clothes.
   C. Fruit.　　D. Books.
5. A. Jason Daniel isn't at home right now.
   B. Jason Daniel doesn't want to answer the phone.
   C. The man can call back later.
   D. The man got the wrong number.

Table Talk

Group Work

Discuss the CPU of your PC with your partner. How does it perform and which kind of CPU do you prefer?

_____

Section Four Translation Skills

科技英语的翻译方法与技巧——科技英语的特点

科技英语用于表达自然科学和工程技术中的相关概念、原理、事实等，强调表达的客观性和真实性，要求语言叙述准确规范、简洁流畅、逻辑性强。科技英语语法就是为了实现这一要求，并且充分体现这一基本特点。科技英语语法的特点如下。

1. 专业词汇多

有些英语词汇在普通英语里和科技英语里的含义在表达时差别很大。

The overrides give you the ability to alter the programmed feed and speed, spindle direction, and rapid traverse motion.

修调（倍率）键用于改变程序中编写的进给速度和主轴转速，主轴转向和快速移动速度。override 在普通英语中有"践踏，代理佣金"的意思，而在数控技术中常常指"倍率，修调"。

又如，apron 在普通英语中是"围裙"的意思，而在车床上翻译成"溜板箱"；engine lathe 就是指普通车床；pocket 有时是"刀套"的意思，有时是"槽，凹处"的意思。这类专业词汇很多，只有大量阅读本专业文献，才能很好的掌握。

2. 被动语态多

据统计，科技英语中的谓语至少有三分之一是被动语态。科技英语中大量使用被动语态，是因为文章需要客观地叙述事理，而不是强调动作的主体。为了强调所论述的客观事物，常把它放在句子的首位。此外，第一、第二人称使用过多，会造成主观臆断的印象。被动语态比主动语态带有更少的主观色彩，这是科技作品所需要的。因此，在科技英语中，凡是在不需要或不可能指出行为主体的场合，或者在需要突出行为客体的场合都需要使用被动语态。

After the layout work completed, machinists perform the necessary machining operations.

规划工作完成以后，机械工就进行必须的加工操作。

Machinists also ensure that the workpiece is being properly lubricated and cooled, because the machining of metal products generates a significant amount of heat.

机械工同时要保证工件被恰当润滑和冷却，因为金属产品加工产生大量切削热。

Dull cutting tools are removed and replaced.

用钝了的刀具要卸下并更换。

3. 定语（从句）多

科技英语中经常需要说明、定义或限制一些概念、条件等，此时需要用定语从句或复杂的限定语来表达。

Lathe is a machine that turns a piece of metal round and round against a sharp tool that gives it shape.

车床是一种相对尖锐的刀具旋转金属件的机床，这种尖锐的刀具使金属件获得所要的形状。在定语从句中还套着一个定语从句 that gives it shape。

Some machinists, often called production machinists, may produce large quantities of one part, especially parts requiring the use of complex operations and great precision.

一些机械工（经常称作制造机械工）可能要大批量地制造某种零件，尤其是那些操作复杂和精度要求高的零件。句中用了过去分词和现在分词做定语。

To repair a broken part, maintenance machinists may refer to blueprints and perform the same machining operations needed to create the original part.

为了修理已损坏的零件，维修机械工要参考图纸，进行与制造新零件所需的相同的机加工操作。

4. 非谓语动词多

英语的每个简单句中，只能用一个谓语动词，如果有几个动词，就必须选出主要动词当谓语，而将其余动作用非谓语动词形式（v.-ing, v.-ed, to v. 3 种形式）表示，才能符合英语的语法要求。

There is a lot of manual intervention required to use a drill press to drill holes.

使用台式钻床钻孔，需要很多人工的干预。这里 required 用过去分词作定语，to use, to drill 都是非谓语动词形式描述动作。又如：

As the bending progresses, the top roll is pressed further down and the radius of the bent workpiece decreases.

随着弯曲进行，进一步顶进，使工件被弯曲的半径减少。

Many computer-controlled machines are partially or totally enclosed, minimizing the exposure of workers to noise, debris, and the lubricants used to cool work pieces during machining.

很多数控机床是全防护或半防护的，最大限度地减少了工人暴露在噪音、切屑碎片和工件冷却润滑液中的可能性。

The process is called die less drawing, as the product being formed without direct contact with a die.

由于产品是在不与模具直接接触的情况下成型，故此工艺为无模拉拨。

5. 复杂长句多

科技文章要求叙述准确，用词严谨，因此一句话里常常包含多个成分。这种复杂且长的句子居科技英语难点之首，阅读翻译时要按汉语习惯加以分析，以短代长，化难为易。

One production machinist, working 8 hours a day, might monitor equipment, replace worn cutting tools, check the accuracy of parts being produced, and perform other tasks on several CNC machines that operate 24 hours a day (light-out manufacturing).

机械制造工，一天工作 8 小时，要监控设备运行，更换用钝的刀具，检查被加工零件的精度，同时在几台 24 小时连续运行（无人值守制造）的数控机床上完成其他工作任务。

Listening

Task I Listen to 2 conversations and choose the best answer.

Conversation 1

1. A. Holiday wear.　　　　　　　B. Sports wear.

C. Summer wear.   D. Casual wear.

2. A. See more samples.   B. Try on some T-shirts.

C. Meet the designers.   D. Place an order.

Conversation 2

3. A. He wanted to check the order number.

B. He wanted to order some computers.

C. He wanted to report on a problem.

D. He wanted to see the secretary.

4. A. They got a wrong order number from the caller.

B. They failed to deliver the computer on time.

C. They couldn't find the order form.

D. They made a wrong deliver.

5. A. The manager.   B. The salesman.   C. Mr. Peterson.   D. Mary.

Task II Listen to the passage and fill in the blanks with the missing words or phrases.

Welcome to you all. We are pleased to have you here to visit our company.

Today, we will first ___1___ you around our company, and then you will go and see our ___2___ and research centre. The research centre was ___3___ just a year ago. You may ask any questions you have during the visit. We will ___4___ to make your visit comfortable and worthwhile.

Again, I would like to extend a warmest welcome to all of you on behalf of our company, and I hope that you will enjoy your stay here and ___5___.

Extensive Reading

Directions: After reading the passage, you are required to complete the sentences below briefly.

History of Mouse

A mouse is a pointing device that detects two-dimensional motion relative to a surface in computing. This motion is typically translated into the motion of a pointer on a display, which allows for fine control of a graphical user interface. Physically, a mouse consists of an object held in one's hand, with one or more buttons. Mice also feature other elements, such as touch surfaces and "wheels", which enable additional control and dimensional input.

Independently, Douglas Engelbart at the Stanford Research Institute (now SRI International) invented his first mouse prototype in the 1960s with the assistance of his lead engineer William English. They christened the device the mouse as early models had a cord attached to the rear part of the device looking like a tail and generally resembling the common mouse. Engelbart never received any royalties from it, as his employer SRI held the patent, which ran out before it became widely used in personal computers. The invention of the mouse was just a small part of Engelbart's much larger project, aimed at augmenting human intellect via the Augmentation Research Center.

Several other experimental pointing devices developed for Engelbart's ON-Line System (NLS) exploited different body movements. For example, head-mounted devices attached to the chin or nose, but ultimately the mouse won out because of its speed and convenience. The first mouse, a bulky device used two wheels perpendicular to each other, the rotation of each wheel translated into motion along one axis. At the time of the "Mother of All Demos", Englebart's group had been using their second generation, 3-button mouse for about a year.

History of Mouse
Mouse:
  a. __1__ device detects __2__ motion relative to a surface
  b. consists of an object held in one's hand, with one or more __3__, and other elements, such as __4__ and __5__
  c. invented by __6__ in the __7__
  d. the first mouse is a bulky device used two __8__ perpendicular to __9__
  e. Engelbart never received any __10__ form it

Section Five Writing

Questionnaire 调查问卷

调查问卷，就是根据调查目的，制定调查问卷，由被调查者按调查问卷所提的问题和给定的选择答案进行回答的一种专项调查形式。问卷调查是一种常用的专项调查手段，是国际通行的一种专项调查形式，既适用于对广大群众关心的问题进行专项调查，也适用于企业了解顾客对某种产品或者服务的意见和建议进行专项调查。

Sample
Questionnaire 调查问卷
  Hello!
  I am doing a research paper about the Future of Chinese Language. You are so lucky to be chosen as one of my interviewees. Your participation will count a lot to the results of my assignment. Please squeeze a few minutes out of your tight schedule and answer the following ten simple questions for me. I would appreciate your help.
  你好！
  我在做一个关于汉语未来的调查报告。你有幸被选为受试者之一。你的参与直接关系到我作业的成绩。请从你紧张的时间里抽出几分钟，回答一下下面10个简单的问题。我会很感激你的帮助。
  Basic Information 基本情况：
  Age 年龄：20
  Sex (M/ F) 性别：M
  Job 职业：Student
  College 学校名称：Liaoning Mechatronics College

Major 专业: Mould design and making

Please answer the following questions:

请回答下列问题：

1. Do you agree that language is an important branch of culture in some sense?

你是否同意在某种意义上，语言是文化的一个重要分支？

2. How many years have you been studying English (or another foreign language)?

你学英语（或者其他一种外语）多少年了？

3. Do you think the ability of using English or another foreign language can help you with your career in the future? And if the answer is yes, in what ways do you think foreign languages are necessary tools?

你认为英语（或其他外语）的能力能否对你将来的事业有所帮助？如果你的答案是肯定的，你认为在那些方面外语是必要的工具？

4. In 2006, how many books written in Chinese have you read (excluding textbooks and books related to your major)?

2006年你一共读过多少本中文书籍（不包括课本和其他与你专业有关的书）？

5. Do you think it necessary that the Chinese National Ministry of Education requires college students to pass the CET-4? And why?

你认为国家教育部要求大学生通过大学英语四级的规定是否合理？为什么？

6. Do you, or anyone around you, spend a lot of time studying to pass some English proficiency tests, CET-4, CET-6, TEM-4 or TEM-8, etc.?

你，或者你身边的人，有没有花费很多时间努力去通过某些英语水平测试，如大英四级、大英六级、专业四级、专业八级等？

7. Which one, A or B, do you spend more hours on per day?

在下面哪一项上你每天花费的时间更多？

A. Previewing or reviewing the texts of the Intensive English Class, reading English newspapers and magazines to improve your English, or listening to BBC or VOA news to improve your listening English.

预习、复习精读课，阅读英文报刊杂志来提高你的英语水平，或者听BBC，VOA新闻锻炼听力能力；

B. Appreciating Chinese novels or poems, reading newspapers or magazines in Chinese, or writing a diary or something else to express your thoughts.

品味中文小说诗歌，阅读中文版刊杂志，或者记日记，写其他记录自己心情的东西。

8. You are copying something very quickly, e.g., notes, but suddenly you forget how to write a certain Chinese character. How often does this happen to you (always, sometimes, seldom or never)?

你在奋笔疾书，例如，记笔记，但是突然你忘记一个字该怎么写。这种情况发生在你身上的频率是多少？（经常，有时候，几乎没有，从来没有）

9. Do you think that more and more Chinese people are learning English, Chinese could be replaced by other international languages, e.g., English, and if so, why?

你认为如果越来越多的人学英文，中文会不会被其他语言（如英语）代替，你为什么这么认为？

10. What can we do to improve the knowledge of the Chinese language in college students and thus preserve the heritage of the Chinese language? (At least two suggestions, please.)

我们能为提高大学生的中文水平做些什么来保留汉语言的文化遗产？（请提至少两点建议）

调查问卷的结构：

1. 问卷说明

每一份问卷的开头，必须有一段简短的前言，说明研究的目的，指导受试者如何回答，作某些必要的说明，以解除受试者的思想顾虑。

2. 被调查者的基本情况

所谓被调查者的基本情况，主要是指被调查者的一些主要特征。

3. 调查问卷的主体内容

所谓调查问卷的主体内容，就是调查者所要调查的基本内容，这是调查问卷中最重要的部分。由于采用问卷的形式，所以调查问卷的主体内容应主要是根据调查目的，提出调查的问题和可供选择的答案。

调查问卷的主体内容设计的好坏，将直接影响整个专项调查的价值。

调查问卷的主体内容主要包括人们的行为、行为后果、态度、意见、感觉、偏好等。

4. 发出调查问卷的单位名称以及负责人姓名

调查问卷的特点：

1. 目的要明确，重点要突出，内容要简洁，不宜过多。
2. 提问自然，用词准确，通俗易懂，实施方便。
3. 格式整齐美观。
4. 应该给调查对象留有一定余地，即自由发表意见的空间。

Practice

You are required to complete the English Questionnaire based on the information given in Chinese.

说明：假如你是王鹏（中国国籍），去海南度假，于2014年7月8日入住三亚宾馆6001房间，7月15日离开，临走时填写了一份问卷调查表。

内容如下：

1. 对酒店的总体管理感到满意；
2. 对酒店提供的各种服务感到满意；
3. 建议：

1）因酒店位于海边，交通并不方便，周围的商业设施也较少，建议酒店每天能提供班车，方便来海边度假的住店客人去市区购买所需物品；

2）建议酒店与相关公司联系，为住店客人提供租车服务。

QUESTIONNAIRE

To improve the quality of our service, we would be grateful if you could complete the following questionnaire.

Name: _____ Nationality: _____ Room Number: _____

Check-in Date: _____ Check-out Date: _____

Did you receive polite and efficient service when you arrived? _____

Are you satisfied with the room service of our hotel? _____

What's your opinion of our health facilities? _____

Please give your impression of our restaurant service. _____

Have you any other comments to help us make your stay more enjoyable?

_____

_____

_____

_____

Sanya Hotel

Section Six Grammar

动名词 Gerund

动名词是动词 ing 形式的一种，是兼有动词和名词特征的非谓语动词。

一、动名词的时态和语态变化形式：

	主动语态	被动语态
一般式	doing	being done
完成式	having done	having been done

其否定形式是在 v-ing 前加上 not。

二、基本用法

（一）作主语

动名词做主语时谓语动词常用单数。

动名词作主语的几种类型：

1. 直接位于句首做主语。例如：

Reading is an art. 读书是一种艺术。

Climbing mountains is really fun. 爬山真是有趣。

2. 用 it 作形式主语，把动名词（真实主语）置于句尾。例如：

It is no use/no good crying over spilt milk. 覆水难收

It was hard getting on the crowded street car. 上这种拥挤的车真难。

3. 用于"There be"结构中。例如：

There is no saying when he'll come. 很难说他何时回来。

There is no joking about such matters. 对这种事情不是开玩笑。

4. 用于布告形式的省略结构中。例如：

No smoking (=No smoking is allowed (here)). 禁止吸烟

No parking. 禁止停车

5. 动名词的复合结构作主语

当动名词有自己的逻辑主语时，常在前面加上一个名词或代词的所有格，构成动名词的复合结构。例如：

Their coming to help was a great encouragement to us. 他们来帮助我们对我们是很大的鼓励。

动名词作主语与动词不定式作主语的比较：

（1）动名词多用来表示泛指或抽象动作，不定式多用来表示特指或具体动作。例如：

Smoking is not good for health. 吸烟有害健康。

It is not good for you to smoke so much. 吸烟对你来说是不太好。

（2）下列句型常用动名词：It is no use (useless, of little use, no good, dangerous) doing sth.

（3）当句子中的主语和表语都是非限定动词时，要遵循前后一致的原则，在形式上要求统一。例如：

Seeing is believing. / To see is to believe. 眼见为实。

（二）作宾语

1. 在及物动词后做宾语。例如：

After hearing the funny story, all of us couldn't help laughing ear to ear.

听到这个有趣的故事，我们忍不住大笑。

Would you mind ringing me up tomorrow? 请你明天给我打电话好吗？

（1）以下动词后，只能跟动名词做宾语：suggest, finish, avoid, stop, can't help, mind, enjoy, require, practice, miss, escape, pardon, advise, consider, imagine, keep, appreciate, escape, permit 等。例如：

I suggest spending our summer vacation in a seaside town. 我建议在海边的一个小镇度过暑假。

（2）动名词和不定式做宾语的区别

部分动词后面，既可接动词不定式，也可接动名词做宾语，意义不变。如：begin, continue, start, hate, like, love, need, require, want 等。但部分动词后接不定式或动名词时，意义差别较大。例如：

Don't forget to post the letter for me. 表示现在或未来的动作

Have you forgotten meeting her in Beijing Airport? 表示动作已经发生

mean to do 打算做某事

I meant to catch up with the early bus. 我打算坐早班公共汽车。

mean doing 意味着

This means wasting a lot of money. 这意味着浪费许多钱。

try to do 设法尽力做某事

You should try to overcome your shortcomings. 你应该设法克服你的缺点。

try doing 试着做某事

Try working out the physics problem in another way.

尽量用另一种方法做出这道物理题。

stop to do 停下一件事去做另一件事

On the way to the airport, I stopped to buy a paper.

在去机场的路上，我停下来买了份报纸。

stop doing 停止做某事
You'd better stop arguing and do as you are told.
你最好停止争论，按我说的去做。
can't help doing 禁不住
They couldn't help jumping up at the news.
听到这个消息，他们情不自禁地跳了起来。
can't help to do 不能帮助干……
I can't help to make up the room for you. 我不能帮你整理房间了。
go on to do 做不同的事或不同内容的事
He went on to talk about world situation. 他接着谈论了世界局势。
go on doing 继续不停地做某同一件事
We'll go on fighting so long as there is oppression in the world.
只要世界上还有压迫，我们就得继续斗争。

2. 在一些短语和介词后作宾语。如：leave off, put off, give up, look forward to, feel like, have trouble / difficulty(in) doing sth., devote to, get used to, pay attention to, be fond of, be worth 等。例如：

I'm looking forward to your coming next time. 期待您的下次光临。
Thank you for offering me so much help. 谢谢你给了我这么多的帮助。
有些介词可以省略，如：have difficulty (in) doing, have no trouble (in) doing, lose time (in) doing, prevent/stop (from) doing, there is no use (in) doing 等。

3. 下列动词后的动名词虽然是被动意义，但不用被动语态。
Your car needs filling. 你这车要充气了。
This city deserves visiting. 这座城市值得光顾一下。
The problem requires studying carefully . 这个问题需要认真研究。
The trees want watering . 这些树需要浇水了。

（三）作表语

1. 动名词作表语与现在分词作表语的区别：
动名词作表语时相当于名词，说明主语的含义及内容，它与主语是同等关系，主语与表语互换位置不影响句子的基本含义，不可用副词来修饰。例如：
Our duty is serving the people. /Serving the people is our duty. 我们的职责是为人民服务。
现在分词作表语时相当于形容词，说明主语的特征、性质、状态等，主语与表语不可互换位置，可用 very, quite 等副词来修饰。例如：
The situation is encouraging. 形势是值得鼓励的。

2. 动名词作表语与不定式作表语的区别：
动名词作表语时表示比较抽象的一般行为、习惯。例如：
My favorite sport is swimming. 我最喜欢的运动是游泳。
不定式作表语时表示具体的某一次动作，特别是将来的动作。例如：
The first thing for us to do is to improve our pronunciation. 我们要做的第一件事是提高我们的发音。

（四）作定语

动名词作定语往往表示被修饰词的某种用途。例如：
a washing machine=a machine for washing=a machine which is used for washing

walking stick 手杖　　　　　　　　opening speech 开幕词
listening aid 助听器　　　　　　　waiting room 候车室
running water 自来水　　　　　　　developing country 发展中国家
working people 劳动人民　　　　　 sleeping child 熟睡的孩子

Exercises

I. Fill in the blanks with proper forms

1. To make a living, he tried _____ , _____ , and various other things, but he had failed in all. (write, paint)
2. She was praised for _____ the life of the child. (save)
3. She ought to be praised instead of _____ . (criticize).
4. Is there any possibility of our _____ the championship? (win)
5. He came to the party without _____ . (invite)

II. Choose the best answer

1. People appreciate _____ with him because he has a good sense of humor.
   A. to work　　　B. to have worked　　　C. working　　　D. have working
2. If I had remembered _____ the door, the things would not have been stolen。
   A. to lock　　　B. locking　　　C. to have locked　　　D. having locked
3. Your shirt needs _____. You'd better have it done today。
   A. iron　　　B. to iron　　　C. ironing　　　D. being ironed
4. He is very busy _____ his papers. He is far too busy _____ callers.
   A．to write；to receive　　　　　B．writing；to receive
   C．writing；receiving　　　　　　D．to write；for receiving
5.—Why was Fred so upset ? —He isn't used _____ criticized.
   A to being　　　B to be　　　C be　　　D being
6. No one enjoys _____ at.
   A. laughing　　　B. to laugh　　　C. to be laughed　　　D. being laughed
7. Grandma said that she had a lot of trouble _____ your handwriting.
   A. to read　　　B. to see　　　C. reading　　　D. in seeing
8. Anything worth _____ is worthy of _____ well.
   A. doing；being done　　　　　　B. doing；doing
   C. to be done；to be done　　　　D. to be done；being done
9. _____ the news of his father's death, he burst into tears.
   A. After hearing　　　B. On hearing　　　C. While hearing　　　D. Having heard
10. The young trees we planted last week require _____ with great care.
    A. looking after　　　B. to look after　　　C. to be looked after　　　D. taking good care

III. Translate the following sentences.

1. 我最喜欢的运动是游泳。
2. 今天去没有用，他不会在家。
3. 你写完作文了吗？
4. 请原谅我来晚了。
5. 他不声不响地走了进来。

CULTURE TIPS

扫一扫认识硬盘

Unit Three
Office Software

Section One Warming Up

Look and learn

Identify the correct name for each picture.

Excel   PowerPoint   Word   Outlook

1._____   2._____   3._____   4._____

Section Two Real World

Listen and act the following dialogue, which is about the usage of office software.

Do you know how to add pictures to a document?

Mary: Technical support department, what can I do for you?

Michael: Hello, I use Microsoft Word to create a **document**. I want to **add** some pictures

to it. But I do not know how to **operate** it.

Mary: Oh, which kind of Word do you use?

Michael: I use MS Word 2007.

Mary: Ok, first you should place the **cursor** in the document where you want to **insert** pictures, then select the Insert **tab** and **click** the Pictures **icon** in the **illustration** group. As soon as you click the Pictures icon, the Insert Pictures dialog box will appear. Then **select** the picture, **image**, **screenshot**, etc. that you want to add in and click the Insert **button**.

Michael: Oh, I see. The selected pictures have been added at the right **location** in the document. Thank you very much.

Mary: You are welcome. Thanks for calling.

Words & Expressions

document ['dɑːkjumənt]	n.	writing that provides information 文档
add [æd]	v.	increase the quality 增加；补充
operate ['ɑːpəreɪt]	v.	work; be in action 操作
cursor ['kɜːrsə(r)]	n.	光标
insert [ɪnˈsɜːrt]	v.	put or introduce into something 插入；嵌入
tab [tæb]	n.	[计算机]选项卡；标签
click [klɪk]	n.	咔嗒声
	v.	make a clicking or ticking sound 点击
icon ['aɪkɑːn]	n.	a graphic symbol [计算机]图标
illustration [ˌɪləˈstreʃən]	n.	an item of information that is representative of a type 说明；插图
select [sɪˈlɛkt]	v.	pick out, 选择
image ['ɪmɪdʒ]	n.	a visual representation 图像；影像
screenshot ['skriːnʃɑːt]	n.	截屏；屏幕快照
file [faɪl]	n.	a set of related records kept together 档案；文件
button ['bʌtn]	n.	an electrical switch operated by pressing 纽扣；按钮
location [loʊˈkeɪʃn]	n.	a point or extent in space 位置；地点
click ... icon		点击……图标
as soon as ...		一……就……
I see.		我明白了。

Unit Three   Office Software

Proper Names

Technical support department	技术支持部门
Microsoft Word	微软公司的文字处理器应用程序
dialog box	对话框

Find Information

Task I Read the dialogue carefully and then answer the following questions.

1. Where does Mary work?
_____

2. What's Michael's problem?
_____

3. Which kind of Word does Michael use?
_____

4. What will happen if Michael clicks the Pictures icon?
_____

5. Does Michael succeed in adding the picture at the end?
_____

Task II Read the dialogue carefully and then decide whether the following statements are true (T) or false (F).

(　) 1. Michael asks Mary for help by sending her an e-mail.
(　) 2. Michael is very good at operating Microsoft word.
(　) 3. The picture can't be inserted into the MS Word document.
(　) 4. Michael should place the cursor in the right place of the document first.
(　) 5. Mary and Michael are talking about the MS Word 2007.

Words Building

Task I Choose the best answer from the four choices A, B, C and D.

(　) 1. If you add 4 _____ 5, you get 9.
　　　　A. for　　　　　B. with　　　　　C. on　　　　　D. to

(　) 2. When will you tell him the good news?
　　　　I will tell him about it as soon as he_____ back.
　　　　A. comes　　　B. came　　　　　C. will come　　D. is coming

(　) 3. We can move the cursor _____ the screen _____ the mouse.
　　　　A. in; by　　　　B. on; by　　　　C. around; with　　D. around; in

(　) 4. —Will you please show me how to operate the new machine?
　　　—Sure. It is a piece of cake. Now let me tell you _____ to do first.
　　　A. how　　　　　B. what　　　　　C. when　　　　　D. which

(　) 5. I found the new flat _____ in pleasant surroundings.
　　　A. being located　　B. locating　　C. located　　D. location

Task II　Fill in each blank with the proper form of the word given.

1. Life is the source of literary _____ . (create)
2. The _____ of this machine is simple. (operate)
3. That store has a good _____ of furniture. (select)
4. He used photographs as _____ for his talk. (illustrate)
5. At the sight of his _____ on the stage, the hall rang with applause. (appear)

Task III　Match the following English terms with the equivalent Chinese.

A — insert a picture　　　　　　1. (　) 状态栏
B — title bar　　　　　　　　　2. (　) 标题栏
C — word count　　　　　　　　3. (　) 菜单栏
D — status bar　　　　　　　　 4. (　) 对话框
E — cell　　　　　　　　　　　5. (　) 单元格
F — menu bar　　　　　　　　　6. (　) 字体
G — save the document　　　　　7. (　) 保存文件
H — dialog box　　　　　　　　8. (　) 帮助
I — font　　　　　　　　　　　9. (　) 插入图片
J — help　　　　　　　　　　　10. (　) 字数统计

Cheer up Your Ears

Task I　Listen and write down what you've heard. Then read and recite till you can use them fluently.

1. Technical support _____, what can I do for you?
2. I want to use Microsoft Word to create a _____.
3. I do not know how to _____ it.
4. You should _____ the cursor in the first paragraph.
5. Click the Pictures _____, the Insert Pictures dialog box will appear.
6. Select the picture, image or _____ that you need.
7. I _____ what you mean.
8. Please click the Insert _____.
9. The selected pictures have been added at the right _____ in the document.
10. Thanks for _____.

Unit Three Office Software

Task II **Listen and fill in the blanks with the missing words and role-play the conversation with your partner.**

A: Hi, Mary! I know you are good at computer. Could you help me with my computer?
B: Of course. What's ___1___ with your computer?
A: My father has bought me a new ___2___ . But it can't work! I don't know why.
B: Don't worry. Let me ___3___ it. Oh, it certainly can't work. You only have the computer hardware system.
A: What's your meaning?
B: I mean if your computer can work, it not only needs the computer hardware system, but also needs the computer ___4___ .
A: Oh, I see.
B: Did you never know it?
A: You know I'm a computer ___5___ .

Task III **Listen to 5 recorded questions and choose the best answer.**

1. A. Yes, I've got it.   B. Good idea.
   C. No, I've got one.   D. Yes, you can.
2. A. Yes, please.   B. No, you wouldn't.
   C. I'd rather you didn't.   D. Yes, it is.
3. A. Yes, I don't.   B. Sorry, it won't happen again.
   C. No problem.   D. No, I don't.
4. A. Sorry, I have no change.   B. I'm a stranger here too.
   C. 30 dollars.   D. It's a nice city.
5. A. I'm not sure.   B. I have no idea.
   C. English book.   D. OK, right away.

Table Talk

Task I **Complete the following dialogue by translating Chinese into English orally.**

A: What are ___1___ (常用的办公软件)?
B: General Office, editing words and making spreadsheets, like Word, Excel.
A: Does your company use it?
B: Yes. ___2___ (所有的公司都会安装这个软件), for they have to make tables, files.
A: Is there anything else except this?
B: ___3___ (比如说交流工具). QQ, MSN and so on.
A: MSN is now very popular in all the major companies.
B: Well, MSN is more convenient than QQ.
A: I think everybody likes Taobao.
B: Yes. ___4___ （网购很流行）. Cheap and convenient.

A: Do we must install foreign trade management software?
B: There are a lot of files that must ___5___ (被保存) in your management software.

Task II   **Pair-work. Role-play a conversation about consulting how to insert a word art in a document with your partner.**

Situation

Imagine you are a consumer of the office software. You want to consult how to insert a word art in a document to the technical department. Now you are talking with the telephone operator.

Section Three Brighten Your Eyes

Pre-reading questions:
1. Can you tell the differences between hardware and software?
2. Are you familiar with the Microsoft office? Which program of it do you use a lot?

The General Introduction of Office Software

Office Software **evolved** to solve the basic problems that business users have in communication, **calculation**, presentation, and **storage** of information. It is a common software that is used in business office. It is a **collection** of programs to be used by people to help working. It often includes word **processor**, spreadsheet and **presentation** program. The word processor enables you to create a document, **store** it on a disk, **display** it on a screen, **modify** it by entering **commands** and words from the **keyboard**, and print it on a printer. A spreadsheet is a **grid** that organizes data into columns and rows. Spreadsheets make it easy to display information, and people can insert **formulas** to work with the data. A presentation program is used to display information, normally in the form of a **slide** show.

--- Words & Expressions ---

evolve [i'vɑːlv]	v.	undergo development or evolution 发展；进化
calculation [ˌkælkjuˈleɪʃn]	n.	the procedure of calculating 计算
storage [ˈstɔːrɪdʒ]	n.	the act of storing something 保管；贮藏
collection [kəˈlɛkʃən]	n.	a set of things of the same type that have been collected 收集；收藏品
processor [ˈprɑːsesə]	n.	(computer science) the part of a computer that does most of the data processing 处理器

presentation [ˌpriːzen'teɪʃn]	n.	the act of presenting something to sight or view 介绍；报告
store [stɔr]	vt.	keep or lay aside for future use 储存；保存
	n.	商店
display [dɪ'spleɪ]	n.	陈列
	vt.	put sth. on show 显示；表现
modify ['maːdɪfaɪ]	v.	change (a plan, an opinion etc.) slightly 修改
command [kə'mænd]	n.	an authoritative direction to do something 命令；指令
keyboard ['kiːbɔːrd]	n.	键盘
grid [grɪd]	n.	a pattern of regularly spaced horizontal and vertical lines 格子；网格
formula ['fɔːrmjələ]	n.	公式
slide [slaɪd]	n.	幻灯片
be used by ... to do sth.		被某人用来做某事
interact with		与……相互作用，与……相互影响；与……相互配合
enable sb. to do sth.		使某人能够做某事

Find Information

Task I Read the passage carefully and then answer the following questions.

1. What is the office software?
   _____

2. What does the office software include?
   _____

3. What can we do by using the word processor?
   _____

4. What is a spreadsheet?
   _____

5. What is a presentation program used to do?
   _____

Task II Read the passage carefully and then decide whether the following statements are true (T) or false (F).

( ) 1. The office software only has one program.
( ) 2. The word processor makes it easy to display information, and people can insert formulas to work with the data.
( ) 3. The word processor is a presentation software.
( ) 4. If we want to display information in the form of a slide show, we can choose the presentation program to help.

(　　) 5. The spreadsheet includes columns and rows.

Words Building

Task I Translate the following phrases into English.

1. 办公软件_____
2. 文字处理程序_____
3. 创建文档_____
4. 输入命令与字符_____
5. 插入公式_____
6. 处理数据_____
7. 显示信息_____
8. 幻灯片_____

Task II Translate the following sentences into Chinese or English.

1. The office software is a collection of programs to be used by people to help working.
   _____

2. The word processor enables you to create a document, store it on a disk, display it on a screen, modify it by entering commands and words from the keyboard, and print it on a printer.
   _____

3. You can type in the command with a keyboard.
   _____

4. His paintings are on display at the exhibition.
   _____

5. This tag does not show up in the slide show view.
   _____

6. 办公系统软件通常包括文字处理程序、电子表格程序和演示程序。
   _____

7. 电子表格使显示信息变得容易，人们能插入公式来处理数据。
   _____

8. 演示程序通常以幻灯片的形式显示信息。
   _____

9. 你已把那些数据存储到磁盘上了吗？
   _____

10. 展览的绘画作品是不出售的。
    _____

Task III Choose the best answer from the four choices A, B, C and D.

(　　) 1. Tom likes cars. He enjoys _____ model cars of all kinds.

| | A. collect | B. collecting | C. collected | D. collection |

( ) 2. —Are all the names of your class listed here?
   —Yes. All _____ mine.
   A. includes   B. including   C. included   D. include

( ) 3. The team finally _____ in reaching the top of the mountain.
   A. achieved   B. enabled   C. managed   D. succeeded

( ) 4. "Tommy, run! Be quick! The house is on fire!" the mother shouted, with clearly in her voice.
   A. anger   B. rudeness   C. regret   D. panic

( ) 5. It was not until Bob took off his dark glasses that I _____ him.
   A. realized   B. requested   C. recorded   D. recognized

Task IV Fill in the blanks with the proper form of the word given.

1. That is a new _____ of the word. (use)
2. The _____ come up with a solution to the system problem. (program)
3. Life is the source of literary _____ . (create)
4. This _____ can print 40 pages in a minute. (print)
5. It is _____ to make a plan than to carry it out. (easy)

Cheer up Your Ears

Task I Listen and write down what you've heard. Then read and recite till you can use them fluently.

1. The office software is _____ to be used by people to help working.
2. The office software often includes _____ , spreadsheet and presentation program.
3. This program enables you to create a _____ .
4. Have you stored the pictures on the _____ ?
5. Modify it by entering commands and words from the _____ .
6. Do you have enough money to buy a new _____ ?
7. A spreadsheet is a grid that organizes data into _____ .
8. Spreadsheets make it easy to _____ information.
9. People can insert formulas to _____ the data.
10. A _____ is used to display information, normally in the form or a slide show.

Task II Listen to 5 short dialogues and choose the best answer.

1. A. She doesn't like to go to the zoo.
   B. She likes to go to the zoo on weekdays.

    C. She doesn't like to go to the zoo on Sundays.

    D. She likes to go to some other places on Sundays.

2. A. Give a message to Mr. Jones.   B. Wait for Mr. Jones.

   C. Write a note to Mr. Jones.   D. Keep Mr. Jones' note.

3. A. She was sick.

   B. She went to London with her mother.

   C. She went to see her sick mother.

   D. Her mother went to see her yesterday.

4. A. December.   B. January.

   C. February.   D. Both January and February.

5. A. In a restaurant.   B. At home.

   C. In a shop.   D. On a train.

Table Talk

Group Work

Talk about the commonly used software with your partner as much as you know.

Section Four Translation Skills

科技英语的翻译方法与技巧——词汇的翻译

要正确、透彻地理解原文，首先必须弄清每句话的关键词语。因此，对词汇的理解尤为重要。

一、词义的选择

词义的选择是正确理解原文的关键之一。在解释中首先碰到的、并且工作量最大的一个问题就是词义的选择。英语词汇中一词多义和一词多性的现象十分普遍，在专业英语中，词义多变的半技术词汇的使用也十分广泛，这就使得词义的选择变得十分重要。例如，green 一般指"绿色的"，但在 green hands 中它的意思是"无经验的"。在解释时要仔细地理解和选择词义。

词义的理解和选择实际上就是单词在特定句子中的含义的判断和确定。词在特定句子中的确切含义，在很大程度上取决于语言环境。词的含义必须根据词在句子中的作用、词的搭配习惯、词的应用场合等方面的情况予以确定。

1. 区分场合，确定词义

词的多义性也反映在词的应用场合上。特别在各种科技文献中，往往同一个词用在不同的专业中就具有不同的词义。例如：

（1）Three to the power of four is eighty-one.

（2）Energy is the power to do work.
译文分别为
（1）3 的 4 次方是 81。
（2）能量是做功的能力。
2. 根据词类，辨明词义
词是通过在句子中担当一定的成分而起作用的。用法不同，词义往往也不同。例如：
（1）Light cannot pass through opaque object.
（2）The foundry produces light castings.
分别译为
（3）光不能穿过不透明的物体。
（4）这个铸造厂生产小铸件。
3. 联系上下文，活用词义
英语中有不少连词具有多义性，因此，同一连词可引导不同种类的从句。这时连词的词义必须通过上下文关系及整个句子的意思来判断和选择。例如：
Small as atoms are, electrons are still smaller.
原子虽小，但电子更小。
4. 按照习惯，搭配词义
翻译时有必要按照汉语的搭配习惯来处理英语句子中的某些搭配词，这样才能得到通顺的汉语译文。例如：
A large pressure and large current are needed to get a large amount of water power.
为了获得充足的水力，需要高水压的强水流。

二、词义的引申
英、汉两种语言在表达方式上存在着许多差别，翻译时在词典里找不到恰当的词义。如果逐词死译，会使译文生硬，不能确切表达原文含义。这时只有根据上下文和逻辑关系，从一个词或词组的基本意义出发，将词义做一定程度的扩展和外延。词义的引申有以下 3 种不同的类型。

1. 具体化引申
当英语句子中采用含义比较抽象的词，而生硬译出会造成概念模糊时，可采用比较具体的词进行解释。例如：
Alloys belong to a half-way house between mixtures and compounds.
合金是介于混合物和化合物之间的一种中间结构。
2. 技术引申
技术引申的目的主要是使专业英语文献中涉及科学技术概念的词语的译名符合我国科学界的语言规范和习惯。例如：
Experience has shown that the applications of mould coats are generally ineffective in eliminating rat-tail from the casting.
经验表明，应用铸型涂料对消除铸件上的鼠尾缺陷一般不起作用。
3. 修辞性引申
修辞性引申的目的完全是为了使译句的语言流畅，语句通顺。例如：
This machine is simple in design, yet it is efficient in operation.

这台机器的结构简单，但工作效率很高。

三、词性的转换

英语和汉语在结构和表达习惯上都有很大的差别。因此，在英译汉时，有时把英语介词、名词译成汉语动词，有时把英语动词译成汉语名词等。词性转换是指解释时改变原文中某些词的词性，以适应汉语表达习惯。常见的有以下几种：

1. 介词转换为动词

The letter A is commonly used for austenite, F for ferrite, G for graphite.

通常用字母 A 表示奥氏体，字母 F 表示铁素体，字母 G 表示石墨。

2. 名词转换为动词或形容词

一个英语句子只能有一个谓语动词，而汉语中动词使用的较多。汉语中使用动词的场合，英语常用名词、形容词、介词、副词等来表示。在英译汉时，应按照汉语的习惯把它们转译成汉语动词。例如：

The experience proved to be a failure.

实验失败了。

3. 形容词转换成其他词

This metal matrix composite is fairly recent development.

这种金属基复合材料是最近开发出来的。

4. 动词转换为名词

The instrument is characterized by its compactness and portability.

该仪器的特点是结构紧凑、携带方便。

5. 副词转换为其他词

The device is shown schematically in figure 3.

图 3 所示为这种装置的简图。

四、词序的变动

词序对译文的流畅性和明确性有重要影响，英汉两种语言表达方式和造句规则不尽相同，在翻译时不仅要注意句子成分的恰当转换，而且还要注意调整或改变原文的词序。例如：

Presented in the paper are new data on this subject.

论文中提出了关于这一课题的新数据。

Listening

Task 1 Listen to 2 conversations and choose the best answer.

Conversation 1

1. A. The director of the finance department.
   B. The manager of the customer services.
   C. The director of the sales department.
   D. The manager of human resources.

2. A. To ask leave for America.
   B. To buy some products for her.

C. To confirm the meeting with her.

D. To apply a job in Stanford Group.

3. A. Putting off the appointment.

   B. Changing the meeting place.

   C. Visiting the bank the next day.

   D. Calling him back that afternoon.

Conversation 2

4. A. Some shoes are missing.   B. Its delivery is delayed.

   C. The order is cancelled.   D. Some packages are damaged.

5. A. Giving an additional discount.   B. Renewing the contract.

   C. Sending the goods by air.   D. Paying for the delivery.

Task II **Listen to the passage and fill in the blanks with the missing words or phrases.**

When you are starting a small business, you should write a business plan. Writing a business plan is the most important ___1___ . This is how people will think about your business. When you ___2___ support from a bank, the bank will read your plan seriously before it gives you any help. Even if you're starting the business with ___3___ , you will still need to have a written plan to help ___4___ your business. The marketing plan is the important part of a business plan to sell the products or ___5___ .

Extensive Reading

Directions: Questions 1 to 4 are based on the following passage.

Personal computers and the Internet give people new choices about how to spend their time. Some may use this freedom to share less time with certain friends or family members, but new technology will also let them stay in closer touch with those they care most about.

I know this from personal experience. E-mail makes it easy to work at home, which is where I now spend most weekends and evenings. My working hours aren't necessarily much shorter than they once were but I spend fewer of them at the office. This lets me share more time with my young daughter than I might have if she'd been born before electronic mail became such a practical tool. The Internet also makes it easy to share thoughts with a group of friends. Say you do something fun, see a great movie perhaps and there are four or five friends who might want to hear about it. If you call each one, you may tire of telling the story. With E-mail, you just write one note about your experience at your convenience, and address it to all the friends you think might be interested. They can read your message when they have time, and read only as much as they want to. They can reply at their convenience, and you can read what they want to say at your convenience.

E-mail is also an inexpensive way stay in close touch with people who live far away. More than a few parents use E-mail to keep in touch, even daily touch, with their children off at college. We just have to keep in mind that computers and the Internet offer another way of staying in touch. They don't take the place of any of the old ways.

1. The purpose of this passage is to _____.
   A. explain how to use the Internet
   B. describe the writer's joy of keeping up with the latest technology
   C. tell the merits(价值) and usefulness of the Internet
   D. introduce the reader to basic knowledge about personal computers and the Internet
2. The usage of E-mails has made it possible for the writer to _____.
   A. spend less time working
   B. have more free time with his child
   C. work at home on weekends
   D. work in public
3. According to the writer, E-mail has an obvious advantage over the telephone because the former helps one _____.
   A. reach a group of people at one time conveniently
   B. keep one's communication as personal as possible
   C. pass on much more information than the later
   D. get in touch with one's friends faster than the later
4. The best title for this passage is _____.
   A. Computer: New Technological Advances
   B. Internet: New Tool to Maintain Good Friendship
   C. Computers Have Made Life Easier
   D. Internet: a Convenient Tool for Communication

Section Five Writing

Agenda 日程表

日程表是表示按日排定的办事或活动程序的一种文书形式。日程表通常包括标题、时间、地点、具体活动安排等。撰写日程表要注意语言简洁、内容清楚，尤其是时间、地点等。为了节省篇幅，做到言简意赅，文字中的虚词通常省略。

Sample

Agenda of International Public Health Conference
Geneva, Switzerland
Oct. 16-17, 2007

Oct. 16th
9:00 a.m. — 11:30 a.m.	Registration (Lobby, Hilton Hotel)
11:30 a.m. — 1:30 p.m.	Lunch & Rest
1:30 p.m. — 2:00 p.m.	Plenary Session: Opening Speech

2:00 p.m. — 6:00 p.m.	Presentation of Research Papers
6:30 p.m.	Dinner
Oct. 17th	
9:00 a.m. —12:00 noon	Workshop
2:00 p.m. — 4:00 p.m.	Visiting Geneva Medical Treatment Center
5:00 p.m. — 6:00 p.m.	Declaration of the List of the Names of the Awarded Papers
6:30 p.m.	Closing Speech by the President of the Conference
7:00 p.m.	Dinner Party

公共健康国际会议议程
瑞士，日内瓦
2007 年 10 月 16 日至 17 日

10 月 16 日	
上午 9:00 — 11:30	签到（地点：希尔顿大酒店大厅）
上午 11:30 —下午 1:30	午餐、休息
下午 1:30 —2:00	全体会议：致开幕词
下午 2:00 —6:00	宣读学术论文
晚 6:30	晚餐
10 月 17 日	
上午 9:00 — 中午 12:00	专题讨论
下午 2:00 — 4:00	参观日内瓦医疗中心
下午 5:00 — 6:00	宣布获奖论文名单
晚 6:30	大会主席致闭幕词
晚 7:00	晚宴

日程表的常用语

registration　签到
second-floor lobby, International Culture Building　国际文化大厦二层大堂
International Reporting Hall　国际报告厅　　the 6th conference room　第六会议室
welcoming dinner　欢迎晚宴　　　　　　　　buffet breakfast　自助早餐
tea break　茶歇　　　　　　　　　　　　　good-bye dinner　欢送晚宴
opening session　开幕式　　　　　　　　　opening speech　致开幕词
closing speech　闭幕词　　　　　　　　　　general (plenary) session　全体会议
academic conference　学术会议　　　　　　workshop　专题讨论
group discussion　小组讨论　　　　　　　　keynote speech　主题发言
questions and answers　自由提问　　　　　photographing　合影
sign the letter of intent　签订意向书　　　　lunch meeting with ...　午餐约见……
presentation of research papers　宣读学术论文
watching performances　观看演出

meeting with personnel director 与人事部门主管碰头

Practice

Translate the following activity schedule into English.

```
活动安排表
星期一，4月18日
下午 4：00     乘航班 CZ8911 到达北京，由亚洲贸易公司的总裁杨光先生到机场迎接
下午 4：15     乘车去长城宾馆
晚 7：30       总裁杨光先生主持晚宴

星期二，4月19日
上午 9：30     在亚洲贸易公司讨论
下午 2：00     小组讨论
晚 7：00       在老舍茶馆观看演出

星期三，4月20日
上午 9：00     讨论
中午 12：00    签订意向书
下午 1：30     吃北京烤鸭
     3：30    参观故宫
     6：00    乘机去上海
```

Section Six Grammar

Participle 分词

分词就是具有动词及形容词二者特征的词，尤指以 -ing 或 -ed,-d,-t,-en 或 -n 结尾的动词性形容词，具有形容词功能，同时又具有各种动词性特点，如时态、语态、带状语性修饰语的性能及带宾词的性能。分词分为现在分词（doing）和过去分词（done）两种，是一种非谓语动词形式。现在分词和过去分词主要差别在于：现在分词表示"主动和进行"，过去分词表示"被动和完成"（不及物动词的过去分词不表示被动，只表示完成）。分词可以有自己的状语、宾语或逻辑主语等。

一、分词的形式（以 do 为例）

时态＼语态	主动语态	被动语态
一般式	doing	being done
完成式	having done	having been done

注意：否定式在 doing 前加 not 即可。

二、分词的用法

1. 作定语

单个分词作定语置于被修饰词的前面，分词短语作定语，置于被修饰词的后面。现在分词修饰的是发出该动作的名词（即与名词有主谓关系），过去分词修饰承受该动作的名词（即与名词是动宾关系）。例如：

boiling water 沸水（强调进行）　　　　boiled water 开水（强调过去）
falling leaves 正在落下的叶子　　　　　fallen leaves 落叶
developing countries 发展中国家　　　　developed countries 发达国家
a moving film 一部感人的电影　　　　　a moved audience 一个被感动的观众

He is a promising young man.
他是一个有前途的年轻人。
We can see the part of the moon lighted by sunlight.
我们能看到被阳光照到的部分月球。
上句中 lighted by sunlight 做后置定语修饰 moon，此句可改为定语从句：
We can see the part of the moon that is lighted by sunlight.

2. 做表语

分词作表语通常当作形容词来用。现在分词表示主语的性质，而且主语多为物；过去分词表示主语的感受或状态，主语多为人。例如：

This is an interesting book. 这是一本有趣的书。
She is interested in books. 她喜欢读书。
注意：
The cup is broken.
杯子碎了。（broken 是过去分词做表语，表示状态）
The cup was broken by the boy. 杯子被那个小男孩打碎了。
（broken 是被动语态的构成部分，说明的是一个动作）

3. 作状语

分词在句子中作状语，可以表示时间、条件、原因、结果、让步、方式、伴随等。

分词（短语）作状语时，其逻辑主语应与句中主语相一致。当现在分词表示的动作发生在谓语动词之前时，则用现在分词的完成式；当所表示动作与谓语动作同时发生，则用现在分词的一般式；完成或被动关系用过去分词。

Being too old, he couldn't walk that far.
因为上了年纪，他走不了那么远。（原因）
While reading the book, he nodded from time to time.
他读书的时候，不停地点头。（时间）
注：表示时间关系的分词短语有时可由连词 when 或 while 引导。
The teacher stood there surrounded by students.
老师站在那里，身边围着学生。（方式）
Standing on the building, you can see the whole city.
如果你站在楼顶，就能看到整个城市。（条件）

4. 作宾语补足语

在 see, watch, hear, observe, notice, feel, find, glimpse, glance 等感官动词和 look at, listen to 等短语动词以及使役动词 have 等后面，与名词或代词构成复合宾语，作宾语补语的成分，有3种形式即动词原形、现在分词和过去分词。动词原形表主动和完成，现在分词表主动或正在进行，过去分词表被动或完成。

I saw the lady getting on the car. 我看到那位女士正在上车。

I saw the lady get on the car and drive off. 我看到那位女士上了车，开走了。

He had his leg broken from the accident. 他出了事故，腿骨折了。

He had the boy working all day long. 他让那个男孩工作一整天。

5. 分词的独立主格结构

分词短语做状语时，其逻辑主语就是句子的主语，否则就必须在分词前另加上自己的逻辑主语（名词或主格代词），这种结构叫做独立主格结构或分词的复合结构。这一结构还可以由 with 或 without 引导。如果分词为 being，being 可以省略。例如：

Time permitting, we will finish the task ahead of time.

如果时间允许，我们将提前完成任务。

His glasses broken, he couldn't see the words on the blackboard.

他眼镜坏了，看不清黑板上的字。

Supper finished (=After supper was finished), we started to discuss the picnic.

吃完晚饭，我们开始讨论野餐的事儿。

The football match (being) over, crowds of people poured out into the street.

足球赛结束后，人群涌到了大街上。

The children looked at me, with their eyes opening wide.

孩子们睁大了眼睛盯着我看。

Exercises

I. Change the following verbs into Present Participles.

1. go _____ 2. come_____ 3. run _____
4. leave _____ 5. regret _____ 6. live _____

II. Change the following verbs into Past Participles.

1. drive _____ 2. break_____ 3. hurt _____
4. weigh _____ 5. regret _____ 6. live _____

III. Fill in the blanks with proper words according to the Chinese meaning.

1. 一首令人愉快的歌曲   a _____ (please) song
2. 一位退休工人   a _____ (retire) worker
3. 所给问题的答案   the answers to the questions _____ (give)
4. 没趣的东西   nothing _____ （interest）
5. 第一本写给孩子的书   the first book _____ (write) for children
6. 东升的旭日   the _____ (rise) sun
7. 广泛使用的语言   the commonly _____ (speak) language

8. 令人困惑的观点　　　　　　　a _____ (confuse) argument
9. 疲劳的一天　　　　　　　　　an _____ (exhaust) day
10. 满意的微笑　　　　　　　　　a _____ (satisfy) smile

IV. Choose the best answer to complete each of the following sentences.

1. _____ with the best students, I still have a long way to go.
   A. Having compared　　B. To compare　　C. Compared　　D. Compare
2. The music of the film _____ by him sounds so _____ .
   A. playing, exciting　　　　　　　　B. played, excited
   C. playing, excited　　　　　　　　D. played, exciting
3. _____ against the coming hurricane, they didn't dare leave home.
   A. Warned　　B. Having warned　　C. To warn　　D. Warn
4. In _____ countries, you can't always make yourself _____ by speaking English.
   A. English-speaking, understand　　　B. English-spoken, understand
   C. English-speaking, understood　　　D. English-spoken, understood
5. After _____ the old man, the doctor suggested that he _____ a bad cold.
   A. examining, should catch　　　　B. examined, had caught
   C. examining, had caught　　　　　D. examined, catch
6. _____ , Tom jumped into the river and had a good time in it.
   A. Be a good swimmer　　　　　　B. Being a good swimmer
   C. Having been good swimmer　　　D. To be a good swimmer
7. _____ how to read the new words, I often look them up in the dictionary.
   A. Having not known　　　　　　　B. Not to know
   C. Don't know　　　　　　　　　　D. Not knowing
8. As his parent, you shouldn't have your child _____ such a book.
   A. read　　B. to read　　C. reading　　D. be reading
9. He returned from abroad _____ that his mother had been badly ill.
   A. heard　　　　　　　　　　　　B. having been heard
   C. having phoned　　　　　　　　D. having been phoned
10. _____ more attention, the tree could have grown better.
    A. Given　　B. To give　　C. Giving　　D. Having given
11. _____ from the top of the tower, we can get a beautiful sight of the city.
    A. To see　　B. Seen　　C. Seeing　　D. See
12. All things _____ , the planned trip will have to be called off.
    A. be considered　　　　　　　　B. considering
    C. Having considered　　　　　　D. considered
13. "Can't you read?" Mary said _____ to the notice.
    A. angrily pointing　　　　　　　B. and point angrily
    C. angrily pointed　　　　　　　D. and angrily pointing
14. The book _____ on the table belongs to me.
    A. which it is　　B. lying　　C. is　　D. which is being

15. Excuse me, but it is time to have your temperature ____.

    A. taking   B. to take   C. take   D. taken

扫一扫认识微软办公软件常用组件

Unit Four

Computer Network

Section One Warming Up

Look and learn

Write the correct word or phrase below each picture.

domain name   browser   protocol   website

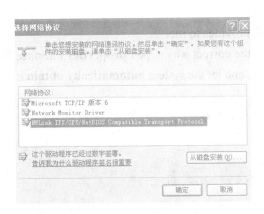

1._____   2._____

3._____

4._____

Section Two Real World

Listen and act the following dialogue which is about the connection to the Internet.

How can I connect to the Internet?

Mary: Hello, Customer Service. Can I help you?

Michael: Hello, I make an **attempt** at connecting to the Internet through ADSL, but I can't access the Internet. What is the problem with it?

Mary: You should first check whether your ADSL **modem** connects to the telephone line in the right way. At the same time you should make sure that your telephone line has been **activated** for **broadband** by your ISP.

Michael: I did. Everything is fine.

Mary: Oh, it requires a **username** and password to connect to the Internet. Do you fill in the correct information?

Michael: Yes, they are the same as those the ISP provides to me.

Mary: Then you should try to **configure** it in the correct way. Open **properties** interface for TCP/IP in your **adapter** properties page, and let the system **automatically obtain** its IP and DNS address.

Michael: I understand. I will get down to do that. Thanks.

Mary: Thank you for calling.

Words & Expressions

attempt [ə'tɛmpt]   vt.   试图
                     n.   activity intended to do or

			accomplish something 尝试
modem	['məʊdem]	n.	electronic equipment consisting of a device used to connect computers by a telephone line 调制解调器
activate	['æktɪveɪt]	vt.	make active or more active 使活动；激活
broadband	['brɔːdbænd]	n.	宽带
username	['juːzəneɪm]	n.	[计] 用户名
configure	[kən'fɪɡə(r)]	v.	set up for a particular purpose 配置；设定
property	['prɒpəti]	n.	a construct whereby objects or individuals can be distinguished 属性；财产
adapter	[ə'dæptə]	n.	适配器；改编者
automatically	[ˌɔːtə'mætɪklɪ]	adv.	in a mechanical manner 自动地
obtain	[əb'teɪn]	vt.	come into possession of 获得
at the same time			同时
make sure			确保
get down to do...			认真处理，对待

Proper Names

ADSL	非对称数字用户线路 (Asymmetrical Digital Subscriber Line)
ISP	互联网服务提供商 (Internet Service Provider)
Property interface	属性界面
TCP/IP	传输控制协议 / 互联网协议 (Transmission Control Protocol / Internet Protocol)
DNS	域名系统 (Domain Name System)

Find Information

Task I Read the dialogue carefully and then answer the following questions.

1. Where does Mary work?

2. Why does Michael call Mary?

3. What should Michael do first?

4. Who provides username and password to connect to the Internet?

5. What should Michael let the system obtain automatically?

Task II Read the dialogue carefully and then decide whether the following statements are true (T) or false (F).

( ) 1. Michael makes an attempt to connect the Internet through ADSL, but he can't access the Internet.

( ) 2. Michael can't access the Internet because his telephone line hasn't been activated for broadband by his ISP.

( ) 3. Michael fills in the wrong username and password.

( ) 4. After Mary's explanation, Michael knows how to let the system automatically obtain its IP and DNS address.

( ) 5. Michael learns how to solve the problem at last.

Words Building

Task I Choose the best answer from the four choices A, B, C and D.

( ) 1. The international law of the Sea Conference is an attempt _____ major differences among countries with conflicting interests.
  A. resolving   B. to resolve
  C. to be resolved   D. resolves

( ) 2. The system has been designed to give students quick and easy _____ to the digital resources of the library.
  A. access   B. passage
  C. way   D. approach

( ) 3. The teacher required that all the students _____ polite to others.
  A. are   B. be
  C. were   D. had been

( ) 4. It's amazing that when I press the red button, the curtains open _____ .
  A. eventually   B. desperately
  C. exactly   D. automatically

( ) 5. In order to buy a house, she had to obtain a _____ from the bank.
  A. finance   B. capital
  C. loan   D. debt

Task II Fill in each blank with the proper form of the word given.

1. She is accused of _____ to cheat the taxman. (attempt)
2. The burglar alarm _____ by movement. (activate)
3. It is easy _____ this software. (configure)
4. A library _____ books through purchase, exchange or by gift. (obtain)
5. Despite my attempts to get him to call me by my Christian name he insisted on _____ me as "Mr. Kennedy". (address)

Unit Four Computer Network

Task III Match the following English terms with the equivalent Chinese.

A——ADSL
B——modem
C——telephone line
D——activate
E——broadband
F——ISP
G——username
H——password
I——properties interface
J——automatically-obtain-its-IP

1. (　) 调制解调器
2. (　) 互联网服务提供商
3. (　) 激活
4. (　) 自动获得 IP
5. (　) 非对称数字用户环线
6. (　) 用户名
7. (　) 属性界面
8. (　) 电话线
9. (　) 宽带
10. (　) 密码

Cheer up Your Ears

Task I Listen and write down what you've heard. Then read and recite till you can use them fluently.

1. I _____ to connect to the Internet through ADSL.
2. I can't _____ the Internet.
3. You should check whether your ADSL _____ connects to the telephone line in the right way.
4. You should make sure that your telephone line has _____ for broadband by your ISP.
5. You will need your _____ for the account.
6. I have filled in the _____ information.
7. You should try to _____ it in the correct way.
8. Open properties interface for TCP/IP in your _____ properties page.
9. Let the system _____ obtain its IP and DNS address.
10. I will _____ down to do that.

Task II Listen and fill in the blanks with the missing words and role-play the conversation with your partner.

A:　__1__ is good!
B: How would you say this?
A: It gives us a lot of __2__.
B: Yes, there are many foreign trade forums, from which we can learn a lot of knowledge.
A: You mean FOB?
B: Yes, I saw yesterday a post titles "how to __3__ unreasonable customers".
A: Interesting.
B: You can play some games, or watch films if __4__.

A: Good choice.

B: Last time I browsed a site, resulting in the virus.

A: There are a lot of __5__.

B: Be careful.

Task III Listen to 5 recorded questions and choose the best answer.

1. A. About 2 hours.   B. It is 2 hours.
   C. About 50 pages.   D. It is very long.
2. A. By gate 12.   B. At the bus station.
   C. Three-thirty.   D. A moment ago.
3. A. It's my pleasure.   B. What about going to the beach?
   C. Yes, I'll go with you. D. No, I don't like it.
4. A. Sorry, I don't know.   B. Yes, I could.
   C. No problem.   D. What is it?
5. A. You have the right.   B. You are welcome.
   C. How much do you plan to save? D. What would you like to know?

Table Talk

Task I Complete the following dialogue by translating Chinese into English orally.

A: Hello, PC Customer Service, __1__（我能为您效劳吗）?

B: Hello, I want to set up a network in my home, how do I choose the best home network setup?

A: The first step towards setting up your network is to find out __2__（你是否拥有所有必需的设备）。

B: There are __3__（网络适配器）in my computers. I also have a __4__（调制解调器）.

A: Then you will have to decide whether you need a wired network or a wireless network. If you only have a stationary computer, then you don't need a wireless network. If you have a number of portable computers, you will need to arrange a wireless network.

A: Oh, I see. There is a computer and a notebook in my home, and I think __5__（无线网络）would be better for me. Thank you so much.

B: You are welcome. Have a nice day.

Task II Pair-work. Role-play a conversation about consulting how to get access to the wireless network.

Situation

Imagine you are having dinner in a fancy restaurant. You want to get access to the wireless network. Now you are talking with the waiter.

Section Three Brighten Your Eyes

Pre-reading questions:

1. How do you connect to the internet?
2. What do you know about the broadband?

The General Introduction of Broadband

Broadband is the name given to any fast, **permanent** Internet connection. It can be delivered by **cable**, **satellite**, mobile phone, **fibre optics** and ADSL.

The difference between **dial-up** and broadband is like the difference between a country **lane** and a motorway. The broadband replaces a single band with separate bands for **uploading**, downloading data and voice. ADSL comes from your **local** phone exchange **via** the old **copper** wires. Your ISP will test and activate the line for you. Some cable TV companies offer a cable internet connection via their existing wiring. Fibre-optics are tiny wires that transmit **pulses** of light. Since light is the fastest thing in the universe, they're much faster than ADSL cables. This means you can get internet speeds up to five times faster than the fastest ADSL connections. Mobile is the newest form of the broadband. A small USB **device** or a data card enables you to access the Internet wherever there is a mobile phone **signal**.

Words & Expressions

permanent ['pɜːmənənt]	adj.	lasting or intended to last for a long time or forever 永久的
cable ['keɪbl]	n.	a set of wires which carry telephone messages, television signals, etc. 电缆
satellite ['sætəlaɪt]	n.	a man-made object intended to move around the Earth, moon, etc. for some purpose 人造卫星
fibre optics ['faɪbə'ɔptɪks]	n.	光纤
lane [leɪn]	n.	a narrow way or road 小路；小巷
upload [ˌʌp'loʊd]	vt.	transfer a file or program to a central computer from a smaller computer or a computer at a remote location 上传；上载
local ['loʊkl]	adj.	belonging to a particular place or district 地方性的；当地的；局部的
via ['vaɪə]	prep.	by means of 经由；通过

copper ['kɒpə(r)]	n.	a metal with a color between brown and red 铜；铜币
pulse [pʌls]	n.	脉冲
device [dɪ'vaɪs]	n.	a piece of equipment intended for a particular purpose 装置；设备
signal ['sɪgnəl]	n.	a sound, image, or message sent by waves, as in radio or television 信号；标志
dial-up	n.	[计]拨号（上网）的
replace…with…		用……代替
enable sb to do sth.		使某人能够做某事

Find Information

Task I Read the passage carefully and then answer the following questions.

1. What is broadband?

2. In which ways can broadband signals be delivered?

3. Who will test and activate the phone line for you?

4. Are cable TV companies capable of offering an internet connection?

5. What is the newest form of the broadband?

Task II Read the passage carefully and then decide whether the following statements are true (T) or false (F).

( ) 1. Broadband can't be delivered by satellite.
( ) 2. Dial-up is like driving in a motorway.
( ) 3. Cable TV companies offer a cable internet connection via mobile phone signal.
( ) 4. A small USB device or a data card enables you to access the Internet wherever there is a mobile phone signal.
( ) 5. Dial-up is not the newest way to connect to the Internet.

Words Building

Task I Translate the following phrases into English.

1. 互联网连接_____
2. 手机信号_____
3. 拨号上网_____

4. 乡间小路_____
5. 本地电话交换系统_____
6. 有线电视_____
7. USB 设备_____
8. 数据卡_____

Task II Translate the following sentences into Chinese or English.

1. The difference between dial-up and broadband is like the difference between a country lane and a motorway.
   _____

2. Sometimes it takes several hours to download a small file.
   _____

3. Local calls are a quarter from a public phone.
   _____

4. Some cable TV companies offer a cable internet connection via their existing wiring.
   _____

5. The strength of the phone signal is very poor in this area.
   _____

6. 宽带是指那些快速的、持续的互联网连接。
   _____

7. 可以通过电缆、卫星、移动电话或者是 ADSL 线路传输（宽带）信号。
   _____

8. 你的互联网服务提供商会为你测试和激活线路。
   _____

9. 移动连接是最新的宽带连接方式。
   _____

10. 一个小的 USB 设备或者是数据卡能让你访问互联网。
    _____

Task III Choose the best answer from the four choices A, B, C and D.

( ) 1. The first textbook _____ for teaching English as a foreign language came out in the 16th century.
   A. writing   B. written
   C. to write   D. to be written

( ) 2. Young _____ he is, he has proved to be an able salesman.
   A. that   B. who
   C. as   D. which

( ) 3. I don't doubt _____ the stock market will recover from the economic crisis.
   A. if   B. what

              C. that   D. which
( ) 4. In our company, great changes _____ since the new manager came.
       A. took place   B. take place
       C. will have taken place   D. have taken place
( ) 5. News came from the sales manager _____ the new product had been selling well
       in the local market for three months.
       A. whose   B. what
       C. which   D. that

Task IV Fill in each blank with the proper form of the word given.

1. He is thinking of living here _____. (permanent)
2. The _____ between who you are and who you want to be is what you do. (different)
3. Sarah sat down and _____ a number. (dial)
4. When dad finished _____ we discovered we used no firewall. (upload)
5. This is all _____ grown produce. (local)

Cheer up Your Ears

Task I Listen and write down what you've heard. Then read and recite till you can use them fluently.

1. Broadband is the name given to any _____ Internet connection.
2. Broadband can be delivered by _____ and ADSL.
3. The difference between dial-up and broadband is like the difference between _____ and _____.
4. The broadband replaces a single band with separate bands for _____ and voice.
5. ADSL comes from your local phone exchange _____ the old copper wires.
6. Some cable TV companies offer a cable internet connection via their _____.
7. Your ISP will test and _____ for you.
8. Mobile is the _____ form of the broadband.
9. A small _____ or a data card enables you to access the Internet.
10. The strength of the _____ is very poor in this area.

Task II Listen to 5 short dialogues and choose the best answer.

1. A. She can not read.   B. Nobody can tell.
   C. She has no watch.   D. She is too sorry.
2. A. He's a boat builder.   B. He smokes a pipe.
   C. He paints watercolors.   D. He's a plumber.
3. A. She goes to bed early.   B. She is single.
   C. She is not tired.   D. She is tired.

4. A. Return some books. B. Go home.
   C. Read Matthew's book. D. Leave class early.
5. A. Everyday. B. Weekends.
   C. Weekdays. D. Seldom.

Table Talk

Group Work

Talk about the basic steps of connecting to the internet with your partner as much as possible.

_____

_____

Section Four Translation Skills

科技英语的翻译方法与技巧——科技英语词汇的构成（1）

英语的词汇构成有很多种，基本词汇是不多的，绝大部分属于构成型词汇。这里仅介绍在专业英语中遇到的专业词汇及其构成。目前，各行各业都有一些自己领域的专业词汇，有的是随本专业发展应运而生的，有的是借用公共英语中的词汇，有的是借用外来语言的词汇，有的则是人为构造的词汇。

1. 派生词 (derivation)

这类词汇非常多，它是根据已有的词加上某种前、后缀，或以词根生成，或以构词成分形成的新词。科技英语词汇中有很大一部分来源于拉丁语、希腊语等外来语，有的是直接借用，有的是在它们之上不断创造出新的词汇。这些词汇的构词成分（前、后缀，词根等）较固定，构成新词以后便于读者揣度词义，易于记忆。

（1）前缀

采用前缀构成的单词在电子商务专业英语中占了很大比例，通过下面的实例可以了解这些常用前缀构成的单词。

multi- 多	multimedia 多媒体	multiprocessor 多处理器
multiplex 多路复用	hyper- 超级	hypercard 超级卡片
hypermedia 超媒体	hypertext 超文本	super- 超级
superpipeline 超流水线	supermarket 超级市场	superset 超集
inter- 相互、在……间	interface 接口	interbusiness 公司间
interlock 互联锁	internet 互联网	micro- 微型
microprocessor 微处理器	microkernel 微内核	microcode 微代码
microkid 微机码	tele- 远程的	telephone 电话
telemarketing 电话营销	teleconference 远程会议	

单词前缀还有很多，其构成可以同义而不同源（如拉丁语、希腊语），可以互换，例如：

multi, poly = many	multimedia 多媒体	polytechnic 多种科技的
uni, mono = single	unicode 统一代码	monochrome 单色的；单色画；单色相片

bi, di = twice	bicycle 自行车	diode 二极管
equi, iso = equa	equality 等式	isograph 等线图
simili, homo = same	similarity 相似性	homogeneous 同样的
semi, hemi = half	semiconductor 半导体	hemicycle 半圆形
hyper, super = over	hypertext 超文本	superscalar 超级标量

（2）后缀

后缀是在单词后部加上构词结构，形成新的单词。常用后缀大致可以分为名词后缀、形容词后缀、动词后缀及副词后缀四类。例如：

-able 可能的	programmable 可编程的	portable 便携的
scalable 可升级的	-ware 件（部件）	hardware 硬件
software 软件	middleware 中间件	groupware 组件
-ity 性质	reliability 可靠性	availability 可用性
integrity 完整性	confidentiality 保密性	-ify 使…化
Webifying 网络化	clientifying 客户化	-ward(s) 方向
downwards 向下	forward 向前	-ee 动作承受者
employee 雇员	examinee 应试者	

2. 复合词 (compound)

复合词是科技英语中另一大类词汇，其组成面广，通常分为复合名词、复合形容词、复合动词等。复合词通常以连词符"-"连接单词构成，或者采用短语构成。有的复合词进一步发展，去掉了连词符，并经过缩略成为另一词类，即混成词。复合词的实例如下：

-based 基于，以……为基础	Windows-based 以
windows 为基础的	credit-based 基于信誉的
file-based 基于文件的	-centric 以……为中心的
client-centric 以客户为中心的	host-centered 以主机为中心的
-oriented 面向……的	market-oriented 市场导向的
process-oriented 面向进程的	user-oriented 面向用户的
-free 自由的，无关的	tax-free 免税的
jumper-free 无跳线的	paper-free 无纸的
charge-free 免费的	info- 信息，与信息有关的
info-sec 信息安全	info-world 信息世界
info-channel 信息通道	info-tree 信息树
e- 电子的	e-commerce 电子商务
e-business 电子商业	e-logistics 电子物流
e-book 电子书	

然而，必须注意，复合词并非随意可以构造，否则会形成一种非正常的英语句子结构。虽然上述例子给出了多个连接单词组成的复合词，但并不提倡这种冗长的复合方式。对于多个单词有非连线形式，要注意其顺序和主要针对对象。

随着词汇的专用化，复合词中间的连接符被省略掉，形成了一个单词。例如：

dotcom 互联网公司	stakeholder 股东	marketplace 市场
website 网页	clickthrough 点击	webcast 网络造型

online	在线	offline	离线	login	登录
logout	注销	popup	弹出	onboard	在船（车）上

此外，还应当注意，有连字符的复合词和不加字符的词意是不同的，必须通过上下文推断。例如，force-feed 强迫接收，而 force feed 则为加润滑油。

3. 混成词 (blending)

混成词不论在公共英语还是科技英语中都大量出现，也有人将它们称为缩合词（与缩略词区别）、融会词。它们多是名词，也有地方将两个单词的前部并接、前后并接或者将一个单词前部与另一个词拼接构成新的词汇，例如：

intermediary (inter + media + ary) 中介
sysop (system + operator) 系统操作员
codec (coder + decoder) 编码译码器
compuser (computer + user) 计算机用户
transceiver (transmitter + receiver) 收发机
syscall (system + call) 系统调用
calputer (calculator + computer) 计算器式计算机
infomercial (information + merchandise + ial) 专题广告片；电视导购节目
brouter (bridge + router) 路由网桥

Listening

Task I Listen to 2 conversations and choose the best answer.

Conversation 1

1. A. The design.　　　　　　　　　B. The price.
   C. The colour.　　　　　　　　　D. The size.
2. A. Black.　　　　　　　　　　　B. Brown.
   C. White.　　　　　　　　　　　D. Gray.
3. A. The discount.　　　　　　　　B. The weight.
   C. The quantity.　　　　　　　　D. The distance.

Conversation 2

4. A. The screen has gone black.　　B. The keyboard doesn't work.
   C. The connections are broken.　　D. The power supply is off.
5. A. Start the computer again.　　　B. Change the mouse.
   C. Check the connections.　　　　D. Replace the keyboard.

Task II Listen to the passage and fill in the blanks with the missing words or phrases.

John Wilson is the ___1___ of a big company. He is always very busy and travels a lot. He will be away for ___2___ all next week. He will visit several cities on his trip, have meeting with some managers from other companies, see a lot of ___3___ and listen to their reports. Mary is his

_____4_____ . And she won't go with her. She's still in the office, answering John's letters and telephones. Keeping contact with other companies and calling John to let him know about anything important. She'll also be in charge of all the employees in the office. She wants to do a good job because she hopes to be ___5___.

Extensive Reading

Directions: After reading the passage, you will find 5 questions. For each question there are 4 choices marked A, B, C or D. You should mark the correct choice.

We are all busy talking about and using the Internet, but how many of us know the history of the Internet? Many people are surprised when they find that the Internet was set up in the 1960s. At that time, computers were large and expensive. Computer networks didn't work well. If one computer in the network broke down, then the whole network stopped. So a new network system had to be set up. It should be good enough to be used by many different computers. If part of the network was not working, information could be sent through another part. In this way computer network system would keep on working all the time.

At first the Internet was only used by the government, but in the early 1970s, universities, hospitals and banks were allowed to use it, too. However, computers were still very expensive and the Internet was difficult to use. By the start of the 1990s, computers became cheaper and easier to use. Scientists had also developed software that made "surfing"(浏览) the Internet more convenient.

Today it is easy to get on-line and it is said that millions of people use the Internet every day. Sending e-mail is more and more popular among students. The Internet has now become one of the most important parts of people's life.

1. The Internet has a history of more than _____ years.
   A. sixty   B. ten
   C. fifty   D. twenty

2. A new network system was set up to _____ .
   A. make computers cheaper
   B. make itself keep on working all the time
   C. break down the whole network
   D. make computers large and expensive

3. At first the Internet was only used by _____ .
   A. the government   B. universities
   C. hospitals and banks   D. schools

4. _____ made "surfing" the Internet more convenient.
   A. Computers   B. Scientists

C. Software   D. Information

5. Which of the following is true?

   A. In the 1960s, computer networks worked well.

   B. In the early 1970s, the Internet was easy to use.

   C. Sending e-mail is now more popular among students than before.

   D. Today it's still not easy to get on-line.

Section Five Writing

Instructions 说明书

说明书主要用来说明产品的性能、特点、用途和使用方法及注意事项等，一般由标题（包括副标题）和正文两大部分组成。标题在说明书中起引导作用。正文包括产品的操作过程、主要功能及产品的维护、保养等。为了醒目，说明书中常列有小标题。为了便于用户理解，说明书的文字要浅显易懂、简单明了，注意科学性和逻辑性，句子结构简单，多用祈使句、简单句。

Sample

Maintenance of Rice Cooker

1. Do not use the cooker on a wet floor, unstable place, or near the gas table.

2. Always keep the flange part and the bottom of the inner pot, the heating plate, the lid, the body and the sensor, etc. clean so that they do not accumulate rice grain or other debris.

3. Do not use the inner pot with other heating appliances such as a gas table.

4. Do not place a wiping cloth, paper, and the like on the lid during use.

5. Never wash the body with water. It may cause electric leakage or problems to the cooker.

6. Do not heat an empty pot.

7. Do not use the cooker to boil water.

8. Do not use the cooker where water can reach.

电饭煲的保养

1. 切勿在潮湿的地板上、不稳定的地方或者煤气炉附近使用。

2. 必须经常保持内锅突起部及底面、加热板、外盖、主机、感应器等部件干净，保证上面没有饭粒或其他杂物。

3. 请勿在其他加热上使用内锅，如煤气炉上。

4. 在使用中，勿将抹布或纸张等放在外盖上。

5. 切勿用水清洗主机，否则会引起漏电或其他故障。

6. 内锅内无饭时，请勿加热内锅。

7. 切勿用该电饭煲煮开水。

8. 请勿在水能溅到的地方使用本电饭煲。

说明书常用语

1. 名词部分

Operation Manual　操作手册
Key Ingredients　主要成分
Warnings　注意事项
Specifications　性能
Date of Manufacture　制造日期
Shelf Life　上架期/保质期
Properties　性质
Action　功能
Indications　适应症/主治
Directions　用法
Administration and Dosage　用法与剂量
Caution / Note　注意/注意事项
Suitable crowd　适宜人群
Net Contents　净含量
Quality Standard　质量标准
Ratification Number　批准文号
Compositions & Contents　功效成分及含量
Quality Ensured Period / Expiration　保质期
Composition and Forms of Issue　成分和包装

2. 动词部分

press　按（下）
pull　拉
push　推
pause　暂停
stop　停
eject　跳出；送出
play　播放
record　录音
rewind　倒带
fast forward　快速前进
reset　复位
zoom　放大
memory　记忆
store　储存
search　搜索
standby　待用状态
To Use　使用方法
To Store/Storage　保存方法

3. 缩写字母及全称

MAX (maximum)　最高音量
MIN (minimum)　最低音量
FM (frequency modulation)　调频
SW (short wave)　短波
MW (medium wave)　中波
CD (compact disc)　光盘
VCD (video compact disc)　视频激光唱盘
SP (speaker)　扬声器
DVD (Digital Video Disk)　数字影像光盘
R (right)　右喇叭
I/O (input/output)　输入；输出
L (left)　左喇叭
MIC (microphone)　麦克风
VOL- (volume minus)　调小音量
PROG + (program plus)　上一个频道
VOL+ (volume plus)　放大音量
PROG - (program minus)　下一个频道

4. 终端输入输出

Video in　视频输入
Video out　视频输出
Audio in　音频输入
Audio out　视频输出
line in　线接入
line out　线接出
on line　线路接通
off line　线未接通
adapter　接合器；接头

5. 电源指示灯

power　电源开关
on　（电源）开

off （电源）关   switch 开关
indicator 指示器；指示灯   voltage 电压
socket 插座   plug 插头

6. 常用句型

Open the package only when ready to use. 准备使用时才打开包装。

Keep out of reach of children. 放在远离儿童的地方。

Avoid contact with the eyes. If product gets into eyes, rinse thoroughly with water. 避免进入眼睛，如不慎渗进眼内，应以清水彻底冲洗。

Keep away from fire. 切勿近火。

Do not touch… 不要触摸……

Take care not to…注意不要……

Store in a vertical position. 垂直存放。

Avoid high temperature and direct sunlight. 避免高温及阳光直射。

Practice

Translate the following instructions into Chinese.

Shark Oil

Major Raw Material: shark oil

Active Ingredient: DHA 250-300 mg, EPA 60-90 mg per 1000 mg

Health Benefit: Improve memory, regulates blood lipids

Dosage: To be taken orally, initially 2 capsules, 2 times per day. For prolonged use, 1-2 capsules, 1 time per day.

Storage: To be stored in a cool, dry place

Shelf Life: 2 years

Section Six Grammar

Attributive clause I 定语从句（一）

定语从句在复合句中做定语，修饰主句中某一名词、代词或整个主句。被定语从句所修饰的词叫先行词，定语从句通常放在先行词后，由"关系词"引导。例如：

I know the boy who is playing the guitar.

我认识那个正在弹吉他的男孩。

The book that is on the desk belongs to my father.

桌上的那本书是我爸的。

引导定语从句的关系词分为关系代词和关系副词两种。

关系代词：who, which, that, whom, whose（在从句中充当主语、宾语或定语）

关系副词：when, where, why　（在定语从句中充当状语）

一个定语从句究竟应该用关系代词还是关系副词来引导是由先行词在定语从句中充当的

成分而定的。具体用法见如下讲解。

1. 当关系词在定语从句中充当主语、宾语时，要用关系代词 who, which, that, whom。

who 用于先行词是表示人的词；

which 用于先行词是表示物的词；

that 则是先行词表人表示物都可用；

whom 要求先行词是表示人的词且其在定语从句中必须充当宾语时才可用。

注意：当关系代词在定语从句中充当宾语时可省，但充当主语时不可省。例如：

（1）Do you know the girl who/that is standing under the tree?

此句中的关系词在定语从句中充当主语且先行词是表示人的词。所以有两个答案：who 或 that，且不能省略。

此句译为：你认识那个站在树下的女孩吗？

（2）This is the book which/that / × I bought yesterday.

此句中的关系词在定语从句中充当宾语且先行词是表示物的词。所以有 3 个答案，即 which, that 或不填，因为宾语可省。

此句应译为：这就是我昨天买的那本书。

（3）He is the man who/whom/that/ × you talked about yesterday.

此句中的关系词在定语从句中充当宾语且先行词是表人的词。所以有 4 个答案，即 Who, whom, that 或不填，因为宾语可省。

此句译为他就是你们昨天谈论的男人。

※ 例句 3 还可把介词提到关系词前。但此时，该句只有一种写法，即

He is the man about whom you talked yesterday.

因为此句的先行词是人，所以介词后必须用宾格，且 whom 不可省。

当先行词是物时，介词后通常加 which。例如：

This is the book which/that/ × I'm interested in.

如果把介词提前，则该句应写为：This is the book in which I'm interested.

2. 当关系词在定语从句中充当状语时，用关系副词 when，where 或 why。

其中，先行词是表时间的词时用 when 引导；先行词是表地点的词时用 where 引导；

而当先行词是原因 reason 一词时用 why 引导。

另外，关系副词通常可以换用介词 + which 的形式。其中，

why = for which;

when = 表时间的介词 + which;

where = 表地点的介词 + which 例如：

（1）He can still remember the day _____ I joined the army.

此句中关系词充当状语且先行词是表时间的词，所以用 when；又因为固定词组 on that day，所以，也可用 on which。即此句中，when = on which。此句应译为：他仍旧记得他入伍的那一天。

（2）That's the room _____ he lived last year.

此句中关系词充当状语且先行词是表地点的词，所以用 where；又因为固定用法 live in the room，所以也可用 in which。此句应译为：那就是他去年住的房间。

（3）I don't know the reason _____ he didn't come this morning.

此句中关系词充当状语且先行词是 reason，所以用 why 或 for which。

此句应译为：我不知道他今天早上没来的原因。

3. whose 对先行词没有要求，但其在定语从句中充当的成分必须是定语，而且先行词与关系词后的名词之间存在一种所属关系时才可用。

（1）The boy whose hair is brown is my brother.

此句中的关系词在定语从句中充当定语且先行词 boy 与关系词后的名词 hair 存在所属关系即"男孩的头发"，所以要用 whose。此句应译为：棕色头发的那个男孩是我的兄弟。

（2）The room whose door is open belongs to Tom.

此句应译为：门开着的那个房间是汤姆的。

综上所述，只要识别出关系词在定语从句中充当的成分，一切问题将会迎刃而解。

完成下列句子，学会正确使用关系词。

① This is the factory _____ makes cars.
② This is the factory _____ he met Lucy two years ago.
③ This is the factory _____ she bought last year.
④ This is the factory _____ my father worked in ten years ago.
⑤ This is the factory in _____ my father worked ten years ago.

Exercises

I. Fill in the blanks with proper relatives.

1. We visited a factory _____ makes toys for children.
2. The doctor _____ you are looking for is in the room.
3. I want the book _____ is on the desk.
4. This is the book _____ you want.
5. He is the boy _____ won the first prize.
6. The shop _____ my sister works is very large.
7. She is the policewoman _____ helped us last week.
8. She is the policewoman _____ you are talking about.
9. She is the policeman about _____ you are talking.
10. Do you know the girl _____ is playing the guitar?

II. Choose the right answer from the four choices.

1. Do you know the girl _____ necklace has been stolen?
   A. her   B. which
   C. that   D. whose
2. When she came back from London, she told us about the schools and teachers _____ she had visited.
   A. who   B. that
   C. which   D. where
3. The notebook _____ cover is red is mine.
   A. who   B. which

C. whose   D. that

4. The man _____ I am going to meet at the station is Professor Smith.
   A. with whom   B. whoever
   C. whom   D. whose

5. Is this the factory _____ you visited the other day?
   A. that   B. where
   C. in which   D. the one

6. Is this the factory _____ he worked ten years ago?
   A. that   B. where
   C. which   D. the one

7. The wolves hid themselves in the places _____ couldn't be found.
   A. that   B. where
   C. in which   D. in that

8. I don't believe the reason _____ he has given for his being late.
   A. why   B. that
   C. how   D. what

9. Is this book _____ I bought for you the day before yesterday?
   A. which   B. that
   C. ×   D. the one

10. I will never forget the days _____ we spent together.
    A. when   B. on which
    C. why   D. that

III. Complete the following conversation with appropriate relative pronouns or adverbs.

A: Shall we invite Jerry to this party?

B: Jerry? Who's he?

A: He's the Italian guy who is staying with Peter's family.

B: Oh, yeah. Is he the one __1__ wallet got stolen when they were in London?

A: That's right. They caught the guy __2__ did it, but he'd already spent all the money __3__ Jerry had bought him.

B: Poor Jerry. Perhaps the party will cheer him up.

A: It might, if we ask the girl __4__ he's been going out with.

B: Who's that?

A: Lucy's her name. She works in that cinema __5__ they show all the foreign films.

B: But will she be free on Thursday evening?

A: Yes, it's her evening off. That's the reason __6__ I suggested Thursday.

B: OK. Who else? What about Nicky and Shirley?

A: Are they the girls __7__ you went to France with?

B: Yes, if they bring their boyfriends, that'll be ten of us. But have you got a room __8__ is big enough? My mother says we can't use our sitting-room because we made too much mess the last time __9__ she let us have a party.

A: It's all right. We've got a basement __10__ we store old furniture. If we clean it up, it'll be fine.

B: Great. Let's go and have a look at it.

 CULTURE TIPS

扫一扫了解 Wi-Fi 使用的安全风险及防范措施

Unit Five

Network Security

Section One Warming Up

Look and learn

Write the correct word below each picture.

spam   firewall   virus   antivirus

1._____   2._____

3._____

4._____

Section Two Real World

Listen and act the following dialogue which is about the network security.

Do you know how to set up a firewall?

Mary: Technical support department, what can I do for you?

Michael: Hello, I just want to know how I can set up a **firewall** for my Windows system.

Mary: You can enable the Internet firewall provided by Windows itself. First go to the Control **Panel**, select Network **Connection**. Then right click on your current connection, such as your LAN or ADSL connection and so on, and select **Properties**. Enter into the advanced tab, enable the Internet Connection firewall. The firewall will work on your computer.

Michael: Should I do something else to protect my computer in the network?

Mary: Of course. This firewall is very **basic** and sometimes **disabled**. You should **install** antivirus software on the computer to fit for your demand. If you need better **security**, it is worth buying some **commercial** firewall products and some other network security software.

Michael: Oh, I see. Thanks a lot.

Mary: You are welcome. Thanks for calling.

Words & Expressions

firewall ['faɪərwɔːl] n. a security system that limits the exposure of a

Unit Five Network Security

			computer or computer network to attack from crackers 防火墙
panel	['pænəl]	n.	a board on which controls or instruments of various kinds are fixed 面板
connection	[kə'nɛkʃən]	n.	a relation between things or events 联系；连接
current	['kɜːrənt]	adj.	of the present time; happening now 现在的
property	['prɑːpərti]	n.	thing or things owned 财产；属性
basic	['besɪk]	adj.	form a base or starting-point 基础的；初级的
disabled	[dɪs'ebəld]	adj.	incapable of functioning 残废的；有缺陷的
install	[ɪn'stɔl]	vt.	set up, ready for use 安装；安置
security	[sə'kjʊrəti]	n.	protection from danger or worry 安全；保证
commercial	[kə'mɜːrʃl]	adj.	商业的
		n.	of or for commerce 商业广告
set up			建立
right click on…			右键单击……
be worth doing…			做……是值得的

Proper Names

Technical support department	技术支持部门
Windows system	Windows 系统
Control Panel	控制面板
Network Connection	网络连接
the advanced tab	高级选项卡
antivirus software	防病毒软件

Find Information

Task I Read the dialogue carefully and then answer the following questions.

1. Where does Mary work?
_____

2. What's Michael's problem?
_____

3. Can Michael set up a firewall by himself?
_____

4. In addition to set up a firewall, what else can Michael do to protect his computer?
_____

5. If Michael needs better security, what should he do?
_____

Task II Read the dialogue carefully and then decide whether the following statements are true (T) or false (F).

(   ) 1. Michael works in the Technical support department.
(   ) 2. Michael knows how to install a firewall.
(   ) 3. Windows system provides Internet firewalls.
(   ) 4. Windows firewall is enough to Michael.
(   ) 5. Michael should buy commercial firewall products if necessary.

Words Building

Task I Choose the best answer from the four choices A, B, C and D.

(   ) 1. "Have great changes taken place in your village?"
"Yes, A new school was _____ in the village last year."
A. held up   B. set up
C. sent up   D. brought up

(   ) 2. His plan was such a good one _____ we all agreed to accept it.
A. as   B. that
C. so   D. and

(   ) 3. No permission has _____ for anybody to enter the building.
A. been given   B. given
C. to give   D. be giving

(   ) 4. Although she's always busy, still she finds time to work _____ charity.
A. on   B. for
C. with   D. as

(   ) 5. To make oneself fit for the job, _____ .
A. it is necessary to be trained in that school
B. a special trained worker is needed
C. we should all go to the training school
D. one needs to be trained in that school

Task II Fill in each blank with the proper form of the word given.

1. The police tried to _____ him with the murder. (connection)
2. This _____ learner's dictionary is very helpful. (advance)
3. We shall be _____ to deal with all sorts of problem. (enable)
4. The document shall be kept in a _____ place. (security)
5. I am afraid your plan is not _____ viable. (commercial)

Task III Match the following English terms with the equivalent Chinese.

A —— firewall   1. (   ) 高级选项

B ——Windows System
C ——Control Panel
D —— Network Connection
E ——current connection
F ——property
G ——the advanced tab
H ——antivirus software
I ——virus
J ——spam

2. (　) 属性
3. (　) 防病毒软件
4. (　) Windows 系统
5. (　) 病毒
6. (　) 控制面板
7. (　) 垃圾邮件
8. (　) 防火墙
9. (　) 当前连接
10. (　) 网络连接

Cheer up Your Ears

Task I Listen and write down what you've heard. Then read and recite till you can use them fluently.

1. _____ support department, what can I do for you?
2. How can I set up a _____ for my Windows system?
3. You can _____ the Internet firewall provided by Windows itself.
4. You should right _____ on your current connection.
5. Enter into the _____ tab.
6. This firewall is very basic and sometimes _____ .
7. You should install _____ software on the computer to fit for your demand.
8. I'll undertake for your _____ .
9. It is worth buying some _____ firewall products.
10. I _____ what you mean.

Task II Listen and fill in each blank with the missing words and discuss the passage with your partner.

Restrict access to a wireless network

It is very easy to restrict(限制) ___1___ to a wireless network and all you need to do is to follow the simple ___2___ given below.

You need to access the wireless router ___3___ page that is generally carried out by typing 192.168.0.1 as the IP address in your web ___4___ address bar.

After you log into the router, you need to enter the security page and enter the key that is the new ___5___ for the wireless access of your network.

Task III Listen to 5 recorded questions and choose the best answer.

1. A. It's fine.
   C. It's October 2nd.
   B. It's Wednesday.
   D. It's National Day.
2. A. Yes, please.
   B. Never mind.

    C. Thanks for doing that.　　　　D. Sorry, I don't know him.
3. A. Thanks, I'd like to.　　　　　　B. It's a pleasure.
    C. That's all right.　　　　　　　D. It's a pity.
4. A. Don't mention it.　　　　　　　B. I'm sorry, I can't.
    C. Not at all.　　　　　　　　　　D. Thanks a lot.
5. A. Yes, I'm feeling very well.　　　B. I feel like swimming.
    C. I've lost my credit card.　　　　D. Don't worry.

Table Talk

Task I Complete the following dialogue by translating Chinese into English orally.

Repairman：What can I do?
Benjamin：___1___（我的系统崩溃了）when I was surfing on the internet.
Repairman：Did you go to any ___2___（非法网站）?
Benjamin：No, but does that matter?
Repairman：Yes, your computer can be easily infected by virus if you do that.
Benjamin：I see. I'd better never try.
Repairman：That's wise.
Benjamin：Do you know ___3___（我的电脑怎么了吗）?
Repairman：One minute. Oh, yes, it was infected by a virus, and you had no ___4___（杀毒软件）.
Benjamin：Is antivirus software necessary for a PC?
Repairman：Of course. You'd better learn something about it.
Benjamin：I'm afraid yes. But ___5___（我原来存在电脑里的数据会怎样）?
Repairman：Don't worry, it should have been protected automatically. And I take an anti-virus software with me. Do you want me to install it now?
Benjamin：Yes, please. I'll really appreciate that.

Task II Pair-work. Role-play a conversation about consulting how to install antivirus software on the computer with your partner.

Situation

Imagine you are a consumer of the 360 antivirus software. You want to consult how to install the software on the computer. Now you are talking with your friend.

Section Three Brighten Your Eyes

Pre-reading questions:

1. Do you use wireless network a lot? How do you protect your computer/ mobile phone from

hacking?

2. Are you familiar with the antivirus software? Which kind of antivirus software do you choose?

The Security Management for Network

Computer security is the process of preventing and **detecting unauthorized** use of your computer. Prevention **measures** help you to stop unauthorized users (also known as "**intruders**") from accessing any part of your computer system. An **effective** network security **strategy** requires identifying threats and then choosing the most effective set of tools to **combat** them.

Security Management for networks is different for all kinds of situations. A small home or an office would only require basic security while large businesses will require high **maintenance** and advanced software and hardware to prevent **malicious** attacks from **hacking** and spamming.

To small homes, every computer connected to the Internet should be protected by a firewall, and that goes double or **triples** for computers on wireless networks. You can either install software firewalls on each computer, or install a hardware firewall on your entire network.

Words & Expressions

detect [dɪˈtekt]	v.	discover or determine the existence 发觉；察觉；探测
unauthorized [ʌnˈɔːθəraɪzd]	adj.	without official authorization 未经授权的；未经批准的
measure [ˈmeʒər]	n.	any maneuver made as part of progress toward a goal 措施
intruder [ɪnˈtruːdər]	n.	someone who intrudes on the privacy or property of another without permission 入侵者；干扰者
effective [ɪˈfektɪv]	adj.	able to accomplish a purpose 有效的
strategy [ˈstrætədʒi]	n.	an elaborate and systematic plan of action 策略
combat [ˈkɑːmbæt]	n.	battle or contend against in or as if in a battle 战斗
	v.	与……斗争
management [ˈmænɪdʒmənt]	n.	control and organization 管理；经营
require [rɪˈkwaɪr]	v.	demand by right 要求；需要
maintenance [ˈmentənəns]	n.	the act of maintaining 维修；维护
malicious [məˈlɪʃəs]	adj.	having the nature of or resulting from malice 怀恶意的；恶毒的

hack [hæk]		vt.	secretly find a way of looking at and/or changing information on Somebody else's computer system without permission 非法侵入（他人计算机系统）
spam [spæm]		n.	unwanted e-mail 垃圾电子邮件
triple ['trɪpəl]		n.	three times as much or as many as sth 3倍
prevent ... from...			阻止
wireless network			无线网络
either ... or...			或者……或者……

Find Information

Task I Read the passage carefully and then answer the following questions.

1. Is the security management for networks the same for all kinds of situations?

2. Which kind of security would a small home or an office require?

3. Which kind of security would large businesses require?

4. Is a firewall necessary for small homes?

5. What can we do to protect the wireless networks?

Task II Read the passage carefully and then decide whether the following statements are true (T) or false (F).

(　) 1. Security Management for networks is the same for all kinds of situations.
(　) 2. A small home network requires advanced security.
(　) 3. Large businesses only require basic security.
(　) 4. Wireless networks need more protection.
(　) 5. Security software and hardware can prevent attacks from outside.

Words Building

Task I Translate the following phrases into English.

1. 安全管理_____
2. 大型商业网络_____
3. 基本安全防护_____
4. 更高维护性_____

5. 先进的软硬件_____
6. 恶意攻击_____
7. 防火墙_____
8. 无线网络_____

Task II Translate the following sentences into Chinese or English.

1. A small home or an office would only require basic security.
   _____
2. Every time I open my email box, I find it's full of spams.
   _____
3. I would advise you to update the software.
   _____
4. One of the best solutions for ensuring Internet safety is a firewall.
   _____
5. Double click on the icon to open the file.
   _____
6. 不同情况下网络安全管理不尽相同。
   _____
7. 大型商业网络则需要性能更好和更先进的软件。
   _____
8. 我们要如何做才能阻止黑客和垃圾邮件的恶意攻击呢?
   _____
9. 点击无线网络并输入你的密码。
   _____
10. 你可以在每台计算机上安装软件防火墙,也可以在整个网络中安装硬件防火墙。
    _____

Task III Choose the best answer from the four choices A, B, C and D.

(   ) 1. Although he did not feel well, he insisted _____ going there together with us.
　　　　A. to　　　　　　　　　　　　B. on
　　　　C. at　　　　　　　　　　　　D. for

(   ) 2. I'll ask Mr. Smith to ring you up _____ he comes back to the office.
　　　　A. when　　　　　　　　　　　B. where
　　　　C. because　　　　　　　　　　D. although

(   ) 3. They regard _____ as their duty to provide the best service for their customers.
　　　　A. this　　　　　　　　　　　 B. what
　　　　C. it　　　　　　　　　　　　 D. that

(   ) 4. Not until the day before yesterday _____ to give a speech at the meeting.
　　　　A. he agreed　　　　　　　　　B. does he agree

C. he agrees   D. did he agree

( ) 5. _____ up at the clock on the wall, the secretary found it was already midnight.
A. Looking   B. Look
C. To look   D. Looked

Task IV Fill in each blank with the proper form of the word given.

1. I want to speak to the sales _____ . (management)
2. How do you tell the _____ between them? (different)
3. Do you think he fills all the _____ for graduation? (require)
4. This site was attacked by a _____ last week. (hack)
5. She put on dark glasses as a _____ against the strong light. (protect)

Cheer up Your Ears

Task I Listen and write down what you've heard. Then read and recite till you can use them fluently.

1. _____ for networks is different from all kinds of situations.
2. A small home or an office would only require _____.
3. Large businesses will require high maintenance and _____.
4. We use antivirus software to prevent malicious attacks from _____.
5. To small homes, every computer _____ should be protected by a firewall.
6. I'll give you _____ for working overtime.
7. I want _____ with mountain view.
8. Click the _____ and type your password.
9. You should _____ on each computer.
10. You can _____ on your entire network.

Task II Listen to 5 short dialogues and choose the best answer.

1. A. In a shoe shop.   B. In a warehouse.
   C. In a department store.   D. In a store for men's clothes.
2. A. It is typical December weather for this region.
   B. It won't snow until December.
   C. There has never been much snow down South.
   D. Such a large amount of snow is unusual for this month.
3. A. He's not sure.   B. He'll go by plane.
   C. He'll go by train.   D. He'll go by bus.
4. A. She must walk five miles.
   B. She must walk five or six blocks.

C. She must walk to the corner.

D. She must walk three blocks.

5. A. It's too soon to go back there again.

B. The mail was sent back to the post office.

C. He doesn't have anything to drop in the mailbox.

D. The post office was closed an hour ago.

Table Talk

Group Work

Talk about the ways you use to protect your network with your partner as much as you can.

_____

_____

Section Four Translation Skills

科技英语的翻译方法与技巧——科技英语词汇的构成（2）

4. 缩略词（shortening）

缩略词是将较长的英语单词取其首字母或者主体构成与原词同义的短单词，或者将组成词汇短语的各个单词的首字母并接为一个大写字母的字符串。随着科技发展，缩略词在文章索引、前序、摘要、文摘、电报、说明书、商标等科技文章中被频繁采用。对计算机专业来说，在程序语句、程序注释、软件文档、文件描述中也采用了大量的缩略词作为标识符、名称等。缩略词的出现方便了印刷、书写、速记以及口语交流等，但同时也增加了阅读和理解的困难。

缩略词出现时，通常采用破折号、引号或者括号将它们的原型单词和组合词一并列出，久而久之，人们对缩略词逐渐接受和认可，作为注释性的后者也就消失了。在通常情况下，缩略词多取自各个组合字（虚词除外）的首部第一、第二字母。缩略词也可能有形同而异义的情况，如果遇到这种情况，翻译应当根据上下文确定词义，并在括号内给出其原型组合词汇。缩略词可以分为以下几种。

（1）压缩和省略

将某些太长、难拼、难记、使用频繁的单词压缩成一个短小的单词，或取其首部、或取其关键音节。例如：

biz = business 商业   lab = laboratory 实验室

math = mathematics 数学   iff = if only if 当切仅当

E-zine = E-magazine 电子杂志 ad = advertisement 广告

Corp = corporation 公司   fiber = fiber optic 光纤

Web = World Wide Web 万维网

（2）缩写（acronym）

将某些词组和单词集合中每个实意单词的第一或者首几个字母重新组合组成为一个新的

词汇，作为专业词汇使用。在应用中形成以下三种类型：

①通常以小写字母出现，并作为常规单词。例如：

radar(radio detecting and ranging) 雷达

laser (light amplification and ranging) 激光

sonar (sound navigation by stimulated emission of radiation) 声呐

spooling (simultaneous peripheral operation online) 假脱机

②以大写字母出现，并作为常规单词。例如：

BASIC (Beginner's All-purpose Symbolic Instruction Code) 初学者通用符号指令代码

ASCII (American Standard Code for Information Interchange) 美国信息交换（用）标准（代）码

COBOL（Common Business Oriented Language） 面向商务的通用语言

NASDAQ (National Association of Securities Dealers Automated Quotations) 美国全国证券交易商协会自动报价表

③以大写字母出现，没有读音音节，仅为字母头缩写。例如：

EDI (Electronic Data Interchange) 电子数据交换

EFT (Electronic Funds Transfer) 电子资金转换

B2B (Business-to-Business) 企业间

B2C (Business-to-Consumer) 企业与消费者间的

B2G (Business-to-Government) 企业与政府间的

C2C (Consumer-to-Consumer) 消费者间的

DNS (Domain Name Service) 域名服务

AOL (America On Line) 美国在线网络

FAQ (Frequently Asked Question) 常见问题解答

LAN (Local Area Network) 局域网

WAN (Wide Area Network) 广域网

ATM (Automated Teller Machine) 自动出纳机

MBPS (Mega Byte Per Second) 每秒兆字节

Mbps (Million Bits Per Second) 每秒兆字位

CIO (Chief Information Official) 高级信息管理人员

HTTP (Hypertext Transfer Protocol) 超文本传输协议

ERP (Enterprise Resource Planning) 企业资源计划

ODBC (Open Database Connectivity) 开放式数据库互连

ISDN (Interprise Services Digital Network) 综合业务数字网

ADSL (Asymmetrical Digital Subscriber Loop) 非对称数字用户环线

CRM (Customer Relationship Management) 客户关系管理

ADE (Application Development Environment) 应用开发环境

DBMS (Database Management System) 数据库管理系统

5. 借用词

借用词一般来自厂商、商标名、产品代号名、发明者名、地名等，也通过将普通公共英语词汇演变成专业词汇而实现。有的则是将原来已经有的词汇赋予新的含义。例如：

Channel conflict 脱媒   platform 平台   cache 高速缓存
Domain 域   woofer 低音喇   hacker 黑客
Virus 病毒（程序、软件）   flag 标志、状态   gateway 网关
Semaphore 信号量   firewall 防火墙   host 主机
Mailbomb 邮件炸弹   scratchpad 便条本   server 服务器
Hub 网络中心   surfing 网上冲浪   Windows 视窗操作系统
Pentium 奔腾处理器   Amazon 亚马逊（网络电子商务公司）
Intel 英特尔公司

在现代科技英语中借用了大量的公共英语词汇、日常生活中的常用词汇，而且以西方特有的幽默和结构讲述科技内容。读者必须在努力扩大自己专业词汇量的同时，掌握和丰富自己的生活词汇，并在阅读和翻译时采用适当的含义。

Listening

Task I Listen to 2 conversations and choose the best answer.

Conversation 1

1. A. He's in his office.   B. He's on holiday.
   C. He's in the meeting room.   D. He's away on business.
2. A. Call him back.   B. Wait for his call.
   C. Send him an email.   D. Visit him in person.

Conversation 2

3. A. The production plan.
   B. Next year's budget.
   C. The sales of the company.
   D. The opening of the new branch.
4. A. Training of new employees.
   B. Development of new products.
   C. Investigation of new markets.
   D. Improvement of the company's sales.
5. A. Alan.   B. The man.
   C. The woman.   D. A guest speaker.

Task II Listen to the passage and fill in each blank with the missing word or phrase.

Ladies and gentlemen, It's a great pleasure to have you visit us today. I'm very happy to have the opportunity to ___1___ our company to you.

The company was established in 1950. We mainly manufacture electronic goods and ___2___ them all over the world. Our sales were about $ 100 million last year, and our business is growing steadily. We have offices in Asia, ___3___ and Europe. We have about 1,000 employees, who are actively working to serve the needs of our ___4___. In order to further develop our overseas market, we need your help to promote our products.

I ___5___ doing business with all of you. Thank you.

Extensive Reading

电子邮件安全使用须知

The followings are some simple ways to protect your computer when using email at home:

- Do not send confidential information in email and don't provide personal information in response to any unknown email or web address you do not trust.
- Do not assume that advice or instructions in email are well intentioned.
- Treat email from people or organizations you don't know with extreme caution. Any attachments received should not be opened until they have been scanned by an up-to-date security software such as Norton Internet Security.
- Install junk email filter so you do not receive useless mail which contains harmful materials.

Task I Record the letters with your partner.

Letters	words
l i f r t e	f
tion in ten	i
struc in tion	i
en fid tial con	c

Task II Match the English with the Chinese Draw lines

1. junk email   a. 安全软件
2. in response to   b. 机密信息
3. security software   c. 对……做出反应
4. confidential information d. 垃圾邮件

Section Five Writing

Signs（标志语）

标志语也叫标识语，是给公众在公共场合看的，表示对大众的一种要求、警告或提示性的文字语言，是日常生活中不可缺少的交际手段，有时甚至起着十分重要的作用。它的特点是内容短小精悍、言简意赅，一般用祈使句、名词、动名词以及短语表达，句末省略标点。

Read and understand the sample signs.

Sample
Admission free 免费入场
Dangerous bend 弯道危险
Caution 注意
Bargain sale 大减价，特价
Beware of pickpockets 谨防扒手
Keep off the grass 请勿践踏草地
Please remove your hat 请脱帽
Close the door after you 请随手关门
Ask inside for details 详情里面询问
Peace of mind from the minute you buy 买着放心
Counter service 柜台服务
No parking 禁止停车

Practice

I. Read the following sings and then match them with their Chinese translations.

(　) 1. No Spitting　　　　　　　　A. 当心碰头
(　) 2. Don't Disturb　　　　　　　B. 禁止拍摄
(　) 3. Mind Your Head　　　　　　C. 切勿挤压
(　) 4. Cash Please　　　　　　　　D. 不准随地吐痰
(　) 5. Do Not Crush　　　　　　　E. 禁止通行
(　) 6. No Thoroughfare　　　　　　F. 电梯修理，请走楼梯
(　) 7. Danger: Keep Out　　　　　G. 请勿打扰
(　) 8. Lift under Repairs, Please Take the Stairs　H. 夜间有事，请按此铃
(　) 9. Cameras Forbidden　　　　　I. 危险，切勿入内
(　) 10. Night Bell　　　　　　　　J. 请付现金

II. Translate the following signs into Chinese.

1. Reception_____
2. Road Ends _____
3. Parking Square _____
4. Lost and Found _____
5. Wet Paint_____
6. Compensation for Damage_____
7. Entrance_____
8. Garbage Prohibited_____
9. Not to be Taken Away_____
10. Free Internet Access_____

Section Six Grammar

Attributive Clause II 定语从句（二）

定语从句分为两种：即限定性定语从句和非限定性定语从句。

限定性定语从句：修饰、限定先行词，是先行词在意义上不可缺少的定语。若省去，意义不完整。其特征是主句和从句之间无逗号隔开，译为："……的……"例如：

The man who/that talked with our teacher yesterday is my father.

昨天跟咱们老师谈话的那个男人是我爸爸。

I like the dress which/that/ × I bought last week.

我喜欢我上周买的那条连衣裙。

非限定性定语从句：对先行词作某些附加说明，若省去，影响不大。其特征是主句和从句之间有逗号隔开。译成汉语时，可分译成另一句话。例如：

I like the dress, which I bought yesterday.

我喜欢这件连衣裙，我昨天才买的。

在非限定性定语从句中，关系词不能由 that 充当，不可省，也不可换。

有时，在非限定性定语从句中，可用 which/ as 来指代主句全句或部分内容。例如：

He didn't pass the exam, which made his parents very angry.（which 指代主句全句）

他考试没及格，这件事使他的父母很生气。

As we all know, China is one of the most densely populated countries in the world.（as 指代主句全句）

众所周知，中国是世界上人口最稠密的国家之一。

* 使用定语从句必须注意的几点如下。

1. 不用 that 的情况

①逗号后，即在非限定性定语从句中。

②介词后。

2. 必须用且只能用 that 的情况

①先行词既有人又有物时。例如：

They are talking about the people and things (that) they saw in America.

他们正在谈论在美国的所见所闻。

②先行词被最高级修饰时。例如：

This is the best film (that) I have ever seen. 这是我看过的最好的电影。

③先行词被序数词修饰时。包括 the last. 例如：

That's the first book (that) she read. 这是她看过的第一本书。

④先行词被 all 所修饰时。例如：

He gave me all the money (that) he saved. 他给了我他攒的所有的钱。

⑤先行词是 all, everything, nothing, anything 等表物的不定代词时。例如：

I'll do all (that) I can to help you. 我会尽所能地帮助你。

⑥先行词被 the very 强调时。例如：That's the very book (that) I want.

⑦句首已有 who/ which 时，为了避免重复。例如：

Read and understand the sample signs.

Sample

Admission free	免费入场
Dangerous bend	弯道危险
Caution	注意
Bargain sale	大减价，特价
Beware of pickpockets	谨防扒手
Keep off the grass	请勿践踏草地
Please remove your hat	请脱帽
Close the door after you	请随手关门
Ask inside for details	详情里面询问
Peace of mind from the minute you buy	买着放心
Counter service	柜台服务
No parking	禁止停车

Practice

I. Read the following sings and then match them with their Chinese translations.

( ) 1. No Spitting　　　　　　　　　　A. 当心碰头
( ) 2. Don't Disturb　　　　　　　　　B. 禁止拍摄
( ) 3. Mind Your Head　　　　　　　　C. 切勿挤压
( ) 4. Cash Please　　　　　　　　　　D. 不准随地吐痰
( ) 5. Do Not Crush　　　　　　　　　E. 禁止通行
( ) 6. No Thoroughfare　　　　　　　　F. 电梯修理，请走楼梯
( ) 7. Danger: Keep Out　　　　　　　G. 请勿打扰
( ) 8. Lift under Repairs, Please Take the Stairs　　H. 夜间有事，请按此铃
( ) 9. Cameras Forbidden　　　　　　I. 危险，切勿入内
( ) 10. Night Bell　　　　　　　　　　J. 请付现金

II. Translate the following signs into Chinese.

1. Reception_____
2. Road Ends _____
3. Parking Square _____
4. Lost and Found _____
5. Wet Paint_____
6. Compensation for Damage_____
7. Entrance_____
8. Garbage Prohibited_____
9. Not to be Taken Away_____
10. Free Internet Access_____

Section Six Grammar

Attributive Clause II 定语从句（二）

定语从句分为两种：即限定性定语从句和非限定性定语从句。

限定性定语从句：修饰、限定先行词，是先行词在意义上不可缺少的定语。若省去，意义不完整。其特征是主句和从句之间无逗号隔开，译为："……的……"例如：

The man who/that talked with our teacher yesterday is my father.

昨天跟咱们老师谈话的那个男人是我爸爸。

I like the dress which/that/ × I bought last week.

我喜欢我上周买的那条连衣裙。

非限定性定语从句：对先行词作某些附加说明，若省去，影响不大。其特征是主句和从句之间有逗号隔开。译成汉语时，可分译成另一句话。例如：

I like the dress, which I bought yesterday.

我喜欢这件连衣裙，我昨天才买的。

在非限定性定语从句中，关系词不能由 that 充当，不可省，也不可换。

有时，在非限定性定语从句中，可用 which/ as 来指代主句全句或部分内容。例如：

He didn't pass the exam, which made his parents very angry.（which 指代主句全句）

他考试没及格，这件事使他的父母很生气。

As we all know, China is one of the most densely populated countries in the world.（as 指代主句全句）

众所周知，中国是世界上人口最稠密的国家之一。

* 使用定语从句必须注意的几点如下。

1. 不用 that 的情况

①逗号后，即在非限定性定语从句中。

②介词后。

2. 必须用且只能用 that 的情况

①先行词既有人又有物时。例如：

They are talking about the people and things (that) they saw in America.

他们正在谈论在美国的所见所闻。

②先行词被最高级修饰时。例如：

This is the best film (that) I have ever seen. 这是我看过的最好的电影。

③先行词被序数词修饰时。包括 the last. 例如：

That's the first book (that) she read. 这是她看过的第一本书。

④先行词被 all 所修饰时。例如：

He gave me all the money (that) he saved. 他给了我他攒的所有的钱。

⑤先行词是 all, everything, nothing, anything 等表物的不定代词时。例如：

I'll do all (that) I can to help you. 我会尽所能地帮助你。

⑥先行词被 the very 强调时。例如：That's the very book (that) I want.

⑦句首已有 who/ which 时，为了避免重复。 例如：

参考答案

Unit One Computer History

Section One Warming Up
1. Bill Gates   2. John Vincent Atanasoff
3. Steve Paul Jobs   4. Douglas C. Engelbart

Section Two Real World

Find information

Task I. 1.The Imitation Game.
       2.It's a biography film about Alan Turing.
       3.It's an apple with one bite missing.
       4.Yes, he did.
       5. He committed suicide.

Task II. 1. F 2. T 3. F 4. F 5. T

Words Building

Task I. 1. B 2. A 3. C 4. A 5. D
Task II. 1. imitate 2. admirable 3. prefer 4. commitment 5. Suicidal
Task III. 1.D 2. F 3. B 4. G 5. A 6. H 7. J 8. E 9. C 10. I

Cheer up Your Ears

Task I. 1. introduced 2. biography; honest 3. character 4. admired; achievement
        5. decode 6. endure; suicide 7. tragic 8. Memory

Task II. 1. computer 2. documents 3. difference 4. memory 5. quality
         6. concerned 7. screen 8. take up 9. amount 10. Pleasure

Task III. 1. C 2. A 3. B 4. D 5. A

Table Talk

1. hard disc 2. capacity 3. RAM 4. out-of-date 5. keep track of

Section Three Brighten Your Eyes

计算机发展史

没有东西能够像计算机一样概括现代生活。不论怎样，计算机渗透到我们社会生活的每一个方面。今日的计算机不只简单地进行计算。要想完全地理解和欣赏计算机给我们生活带来的影响和对未来生活提供的希望，就必须理解它们的进化历史。

1. 第一代计算机（1946～1959年）

第一代计算机以其操作指令是为特定任务特别制定且计算机是为这一特定任务而使用的为特点。每台计算机有不同的叫作机器语言的二进制编码程序来指导其运作。

2. 第二代计算机（1959～1964年）

储存程序和编程语言使得计算机变得灵活，低成本、高产出而最终为商业所用。新兴职业（如程序员、分析师和计算机系统专家）和整个软件产业随着第二代计算机兴起。

3. 第三代计算机（1964～1971年）

随着更多的部件被压缩在了集成芯片上，计算机变得更小。另一类三代计算机的发展也包括使用一个中心程序的计算机操作系统的，它能同时运行多个程序。

4. 第四代计算机（1971年至今）

集成电路之后，计算的发展只有"变小"这一条道路，这也就是说，大规模集成使得成百上千的部件都压缩在一个芯片上。

5. 第五代计算机（现在及未来）

要定义第五代计算机在某种程度上来说很难，因为这一领域仍在起步阶段。现代计算机的某些特质与五代计算机有异曲同工之处。
Find Information
Task I. 1.No, they don't. 2.Machine language.
    3.Software industry. 4.In size. 5.Because it's in its infancy.
Task II. 1. F 2. T 3. F 4. T 5. F
Words Building
Task I. 1. a binary-coded program   2. machine language
    3. stored program   4. the integrated circuit chip
    5. large scale integration   6. in one's infancy
Task III. 1. B 2. D 3. C 4. C 5. C
Task IV. 1. appreciation 2. promising 3. evolution 4. effective 5. entirely
Cheer up Your Ears
Task I. 1. infiltrated 2. appreciate; evolution 3. characterized 4. language
    5. flexibility 6. careers; industry 7. squeezed 8. infancy
Task II. 1. C 2. A 3. D 4. B 5. A
Listening
Task I. 1. C 2. B 3. D 4. A 5. B
Task II. 1. deal with 2. useful 3. effect 4. totally 5. wonderful
Extensive Reading

阿兰·图灵

　　阿兰·麦席森·图灵（1912年6月23日生人，卒于1954年6月7日），是英国的计算机先驱科学家、数学家、逻辑学家、密码专家、哲学家、数学生物学家和马拉松和超长跑运动员。他在计算机科学发展中举足轻重，以图灵机提出了正规化的"算法"和"计算"的概念，而图灵机也被认为是通用计算机的模型。图灵被大众认为是理论计算机科学和人工智能之父。

　　在二战期间，图灵为位于布莱切利公园的政府信号密码学校工作，这也是英国的密码破译中心。曾有一段时间，他负责破解德国海军密码的Hut 8的工作。图灵设计了一系列的技术，设备来破译德军的密码，其中就包括运用炸弹方法，这种方法可以用来寻找Enigma设备的设置。图灵关键作用在截获破译加密信息使盟军在许多重要活动，包括大西洋战役中战胜纳粹；据估计他在布莱切利公园的工作使得欧洲战场至少提前两年结束。

　　图灵在1952年因其同性恋行为遭到起诉，当时这种行为在当时的英国仍被认定是犯罪。他接受雌激素治疗注射(即化学阉割)而免去牢狱之灾。图灵于1954年6月7日，也就是在他42岁生日的前16天，死于氰化物中毒。事后调查显示他死于自杀，而已知证据也表明他死于意外中毒。在2009年，互联网行动之后，英国首相戈登·布朗代表英国政府为"他受到的非人待遇"向公众致歉。英女王伊丽莎白二世在2013年赦免了图灵。

1. male 2. 23 June, 1912 3. 7 June, 1954 4. cyanide poisoning 5. British 6. pioneering computer scientist 7. marathon 8. theoretical computer science 9. artificial intelligence 10. Europe

Section Four Writing

Sanderson Motor Group	
William White	
General Manager	
Address: 12 Nicholson Avenue	Tel: 632574721
Canberra City	Fax: 632574723
Austrilia	
Post Code: ACT 2601	E-mail: William@sanderson.com

参考答案

Unit One Computer History

Section One Warming Up
1. Bill Gates   2. John Vincent Atanasoff
3. Steve Paul Jobs   4. Douglas C. Engelbart

Section Two Real World
Find information
Task I. 1.The Imitation Game.
       2.It's a biography film about Alan Turing.
       3.It's an apple with one bite missing.
       4.Yes, he did.
       5. He committed suicide.
Task II. 1. F 2. T 3. F 4. F 5. T

Words Building
Task I. 1. B 2. A 3. C 4. A 5. D
Task II. 1. imitate 2. admirable 3. prefer 4. commitment 5. Suicidal
Task III. 1.D 2. F 3. B 4. G 5. A 6. H 7. J 8. E 9. C 10. I

Cheer up Your Ears
Task I. 1. introduced 2. biography; honest 3. character 4. admired; achievement
        5. decode 6. endure; suicide 7. tragic 8. Memory
Task II. 1. computer 2. documents 3. difference 4. memory 5. quality
         6. concerned 7. screen 8. take up 9. amount 10. Pleasure
Task III. 1. C 2. A 3. B 4. D 5. A

Table Talk
1. hard disc 2. capacity 3. RAM 4. out-of-date 5. keep track of

Section Three Brighten Your Eyes

计算机发展史

没有东西能够像计算机一样概括现代生活。不论怎样，计算机渗透到我们社会生活的每一个方面。今日的计算机不只简单地进行计算。要想完全地理解和欣赏计算机给我们生活带来的影响和对未来生活提供的希望，就必须理解它们的进化历史。

1. 第一代计算机（1946～1959年）
    第一代计算机以其操作指令是为特定任务特别制定且计算机是为这一特定任务而使用的为特点。每台计算机有不同的叫作机器语言的二进制编码程序来指导其运作。

2. 第二代计算机（1959～1964年）
    储存程序和编程语言使得计算机变得灵活，低成本、高产出而最终为商业所用。新兴职业（如程序员、分析师和计算机系统专家）和整个软件产业随着第二代计算机兴起。

3. 第三代计算机（1964～1971年）
    随着更多的部件被压缩在了集成芯片上，计算机变得更小。另一类三代计算机的发展也包括使用一个中心程序的计算机操作系统的，它能同时运行多个程序。

4. 第四代计算机（1971年至今）
    集成电路之后，计算的发展只有"变小"这一条道路，这也就是说，大规模集成使得成百上千的部件都压缩在一个芯片上。

5. 第五代计算机（现在及未来）
    要定义第五代计算机在某种程度上来说很难，因为这一领域仍在起步阶段。现代计算机的某些特质与五代计算机有异曲同工之处。
Find Information
Task I. 1.No, they don't. 2.Machine language.
　　　　3.Software industry. 4.In size. 5.Because it's in its infancy.
Task II. 1. F 2. T 3. F 4. T 5. F
Words Building
Task I. 1. a binary-coded program　　2. machine language
　　　　3. stored program　　　　　　4. the integrated circuit chip
　　　　5. large scale integration　　　6. in one's infancy
Task III. 1. B 2. D 3. C 4. C 5. C
Task IV. 1. appreciation 2. promising 3. evolution 4. effective 5. entirely
Cheer up Your Ears
Task I. 1. infiltrated 2. appreciate; evolution 3. characterized 4. language
　　　　5. flexibility 6. careers; industry 7. squeezed 8. infancy
Task II. 1. C 2. A 3. D 4. B 5. A
Listening
Task I. 1. C 2. B 3. D 4. A 5. B
Task II. 1. deal with 2. useful 3. effect 4. totally 5. wonderful
Extensive Reading

阿兰·图灵

　　阿兰·麦席森·图灵（1912年6月23日生人，卒于1954年6月7日），是英国的计算机先驱科学家、数学家、逻辑学家、密码专家、哲学家、数学生物学家和马拉松和超长跑运动员。他在计算机科学发展中举足轻重，以图灵机提出了正规化的"算法"和"计算"的概念，而图灵机也被认为是通用计算机的模型。图灵被大众认为是理论计算机科学和人工智能之父。

　　在二战期间，图灵为位于布莱切利公园的政府信号密码学校工作，这也是英国的密码破译中心。曾有一段时间，他负责破解德国海军密码的Hut 8的工作。图灵设计了一系列的技术，设备来破译德军的密码，其中就包括运用炸弹方法，这种方法可以用来寻找Enigma设备的设置。图灵关键作用在截获破译加密信息使盟军在许多重要活动，包括大西洋战役中战胜纳粹；据估计他在布莱切利公园的工作使得欧洲战场至少提前两年结束。

　　图灵在1952年因其同性恋行为遭到起诉，当时这种行为在当时的英国仍被认定是犯罪。他接受雌激素治疗注射(即化学阉割)而免去牢狱之灾。图灵于1954年6月7日，也就是在他42岁生日的前16天，死于氰化物中毒。事后调查显示他死于自杀，而已知证据也表明他死于意外中毒。在2009年，互联网行动之后，英国首相戈登·布朗代表英国政府为"他受到的非人待遇"向公众致歉。英女王伊丽莎白二世在2013年赦免了图灵。

1. male 2. 23 June, 1912 3. 7 June, 1954 4. cyanide poisoning 5. British 6. pioneering computer scientist 7. marathon 8. theoretical computer science 9. artificial intelligence 10. Europe

Section Four Writing

	Sanderson Motor Group	
	William White	
	General Manager	
Address: 12 Nicholson Avenue		Tel: 632574721
Canberra City		Fax: 632574723
Austrilia		
Post Code: ACT 2601		E-mail: William@sanderson.com

Section Five Grammar

I. 1. B 2. D 3. B 4. D 5. B 6. C 7. C 8. D 9. B 10. C
11. A 12. C 13. C 14. C 15. D

II. 1. to say 2. to have 3. for, to, read 4. enough, to lift
5. in, order, to, get 6. to visit 7. to go 8. to have
9. for, to carry 10. to lie

Unit Two Hardware

Section One Warming Up
1.CPU 2.Main board 3.Mouse 4.Printer 5.Keyboard 6.Monitor

Section Two Real World
Find information
Task I. 1.No, they aren't.
2. He can't tell the difference between RAM and ROM.
3. It stands for Random Access Memory.
4. It represents Read Only Memory.
5. ROM.
Task II. 1. T 2. F 3. F 4. T 5. T

Words Building
Task I. 1. A 2. B 3. A 4. A 5. D
Task II.
1. confusing 2. memorable 3. alteration 4. difference 5. erase
Task III. 1. C 2. F 3. B 4. H 5. G 6. E 7. D 8. A 9. J 10. I

Cheer up Your Ears
Task I. 1. components 2. difference 3. confused 4. Random 5. switch
6. essential 7. represent 8. alter 9. journal; erase
Task II. 1. hard drive 2. monitor 3. sound card 4. video card 5. speakers
6. software 7. service 8. charge 9. DVD 10. 8 000
Task III. 1. C 2. A 3. A 4. C 5. C

Table Talk
1. ask you a question 2. dust, dirt and liquids 3. glass cleaners
4. electric shock 5. be conducted to

Section Three Brighten Your Eyes

中央处理器

在一独立使用（非联网状态下）的情况下，计算机的性能绝大部分以3个计算机部件决定，它们分别是中央处理器、随机存储器和显示器决定。

如同大脑一样，中央处理器控制信息并指示其他部件行动。它处理计算机指令执行工作，就和汽车的引擎把油料变成动力一样。中央处理器的性能以其每秒可以执行的指令最大值，即赫兹来表示。1赫兹相当于每秒处理100万个指令。当代的中央处理器相当快速。如果一中央处理器在350赫兹或以上运作，大多数的用户不会体会到额外增加的速率。当然，差别还是有的，只是很微弱，不明显罢了。

大多数现代的中央处理器应用微型处理器，这就意味着他们都在一个集成芯片上面。包含中央处理器的集成芯片上还有可能包含储存器，外围接口和计算机的其他部件；这样的集成器件有多种多样的称呼，如微控制器或系统级芯片。某些计算机使用多核处理器，就是在一芯片上包含两个或多个叫作"核"的中央处理器；在那一情况下，单一的芯片别叫作"套接口"。阵列处理器或向量处理器都有多个平行运作的处理器，没有单一一个部件被认为是核心的。

Find Information
Task I. 1. The CPU, RAM and the computer monitor.
2. It controls information and tells other parts what to do.
3. Megahertz.
4. It means they are contained on a single integrated circuit (IC) chip.
5. It may contain memory, peripheral interfaces, and other components of a computer.

Task II. 1. T 2. F 3. T 4. T 5. F

Words Building
Task I. 1. Central Processing Unit   2. computer performance
3. an integrated circuit (IC) chip   4. peripheral interfaces
5. a multi-core processor

Task III. 1. C 2. A 3. B 4. B 5. D

Task IV. 1. processor 2. appreciate 3. integrated 4. referred 5. determined

Cheer up Your Ears
Task I. 1. determined 2. information 3. processes; processes 4. measured 5. equals 6. incredibly
7. obvious 8. integrated 9. referred 10. considered

Task II. 1. C 2. D 3. B 4. A 5. D

Listening
Task I. 1. C 2. A 3. C 4. B 5. C

Task II. 1. show 2. factory 3. built 4. do our best 5. have a good time

Extensive Reading

鼠标的历史

在计算机在中，鼠标是一种定位设备，用来探测相互对于表面的二维运动。这一运动通常被转换成显示器上的运动指针，这使得用户的图形界面得以有效控制。实际上，鼠标包括有一个或两个按键的可握在手中的物体组成。鼠标也常常包括其它元素，如触摸界面和"滑轮"，这也增加了可控性和多维输入。

斯坦佛研究中心（现在叫作斯坦福国际研究所）的道格拉斯·恩格尔巴特于20世纪60年代，在他的首席工程师比尔·英格里希的协助下，独立发明了他的一个鼠标原型。他们之所以命名这个设备为"鼠标"，是因其早期模型有一根绳子连接其后部，就跟老鼠的尾巴一样。恩格尔巴特从未因鼠标的发明得到任何好处，因为斯坦福国际研究所享有专利权，而这一专利早在个人电脑广泛使用前就过期了。鼠标的发明仅仅是恩格尔巴特宏伟项目中的一小部分，意在借由增强研究中心来增强人类智力。

恩格尔巴特的其它几项实验性指示设备的开发是为了他的在线系统（NLS）能够利用身体的不同部位运动——如，附加到下巴和鼻子的头戴设备—但最终鼠标以其快速性和便捷性取胜。一代鼠标是一个大块的设备，使用垂直于彼此的连个轮子，每个轮子沿着一个轴旋转转化成运动。在"演示之母"时代，恩格尔巴特的团队使用他们的第二代三按键鼠标长达一年。

1. pointing 2. two-dimensional 3. buttons 4. touch surfaces 5. wheels 6. Douglas Engelbart
7. 1960s 8. wheels 9. each other 10. royalties

参考答案

Section Four Writing

QUESTIONNAIRE

To improve the quality of our service, we would be grateful if you would complete the following questionnaire.

Name: Wang Peng Nationality: Chinese Room Number: 6001
Check-in Date: July 8, 2014   Check-out Date: July 15, 2014
Did you receive polite and efficient service when you arrived? Yes
Are you satisfied with the room service of our hotel?   Yes
What's your opinion of our health facilities?   Good
Please give your impression of our restaurant service.   Satisfied
Have you any other comments to help us make your stay more enjoyable?
　　I honestly suggest the hotel provide hotel guests with shuttle bus service to shop necessities in the city, for the hotel is located at seaside with poor traffic to the city and little commercial facilities nearby.
　　I'd also like to suggest that the hotel work with the related companies to offer rental car service for hotel guests.

　　　　　　　　　　　　　　　　　　　　　　　　　　　　Sanya Hotel

Section Five Grammar

I. 1.writing; painting   2.having saved   3.being criticized   4.winning
　5.being invited
II. 1. C 2. A 3. C 3. B 5. A 6. D 7. C 8. A 9. B 10. A
III. 1. My favorite sport is swimming.
　2. It's no use going there today. He won't be in/(can't be in).
　3. Have you finished writing your composition?
　4. Excuse me for being/(coming) late.
　5. He entered the room without making any noise.

Unit Three office software

Section One   Warming Up
1. word 2. excel 3. power point 4. outlook

Section Two   Real World
Find Information
Task Ⅰ :
1. She works in the Technical support department.
2. He doesn't know how to add pictures to a document.
3. He uses MS Word 2007.
4. The insert pictures dialog box will appear.
5. Yes, he does.
Task II: 1.F 2. F 3. F 4. T 5. T
Words Building
Task I: 1.D 2. A 3.B 4.B 5. C
Task II: 1. creation 2.operation 3.selection 4.illustration 5.appearance
Task III: 1.D 2. B 3.F 4. H 5. E 6. I 7. G 8. J 9. A 10. C

参考答案

Cheer up Your Ears
Task I: 1.department 2.document 3.operate 4.place 5.icon
6.screenshot 7.see 8.button 9.location 10.calling
Task II: 1.wrong 2. laptop 3. check 4. software system 5. outsider
Task III: 1. A 2. C 3. B 4. C 5. C

Table Talk
Task I:
1. the common office software
2. All companies will install this software.
3. Like communication tools
4. Online shopping is very popular
5. be saved

Section Three Brighten Your Eyes

办公软件简介

办公软件的发展用来解决企业用户在沟通、计算、演示和信息存储中所遇见的基本问题，是用于商务办公中的常见软件。它是一组程序的集合，用于帮助人们工作。它通常包括文字处理程序、电子表格程序和演示程序。使用文字处理程序，您能够创建文档，并储存在硬盘上，能在屏幕上显示文档，能通过键盘输入命令和字符对文档进行修改，也可以通过打印机打印文档。电子表格是把数据组织在行和列中的表格，它使显示信息变得容易，人们能插入公式来处理数据。演示程序通常用于以幻灯片的形式显示信息。

Find Information
Task I: 1. The office software is a collection of programs to be used by people to help working.
2. It often includes word processor, spreadsheet and presentation program.
3. The word processor enables you to create a document, store it on a disk, display it on a screen, modify it by entering commands and words from the keyboard, and print it on a printer.
4. A spreadsheet is a grid that organizes data into columns and rows.
5. A presentation program is used to display information, normally in the form or a slide show.
Task II: 1. F 2. F 3. F 4. T 5. T

Words Building
Task I:
1. office software   2. word processor
3. create a document   4. enter commands and words
5. insert formulas   6. work with the data
7. display information   8. slide show
Task III: 1. B 2. B 3. D 4. D 5. D
Task IV: 1. usage 2. programmer 3. creation 4. printer 5. easier

Cheer up Your Ears
Task I:
1. a collection of programs   2. word processor
3. document   4. disk
5. keyboard   6. printer
7. columns and rows   8. display
9. work with   10. presentation program

参考答案

Task II: 1. C 2. A 3. C 4. B 5. A

Listening

Task I. 1. D 2. C 3. D 4. B 5. A

Task II. 1. step 2. are looking for 3. your own money 4. develop 5. your services

Extensive Reading

　　个人电脑和互联网给人们如何消磨时间提供了一个新的选择。一些人可能会因为使用网络而与变得很少与朋友或家人共享时光，但新技术将会让他们与那些他们最关心的人保持联系。

　　我是从个人的经历了解到这些的。我现在会在家里度过大部分周末和晚上时间，因为电子邮件使得在家里工作变得更容易。我的工作时间并没有比从前变短，但我在办公室花的时间却更少了。如果我的女儿出生在电子邮件并没有成为这样一个实用的工具的时代，那么我可以陪她的时间要比现在少得多。互联网也使与朋友们交流思想变成易事。假如您做了有趣的事情，看一场很棒的电影，想要分享给四五个朋友。如果您给每一个朋友打电话，您一遍又一遍地讲这个故事可能会感到厌倦。有了电子邮件，您只需要写一个关于您的经历的信息，在您的方便的时候，发送给所有您认为会感兴趣的朋友。当他们有时间时就会阅读您的信息，并且只读他们有兴趣的那一部分。他们可以在方便的时候回复，您可以在方便的时候读他们的留言。

　　电子邮件同时是一种与住在远方的人保持密切联系的廉价的方式。很多家长会通过电子邮件与他们上大学的孩子保持联系，甚至是每天都联系。我们只需要记住，电脑和互联网提供了另一种保持联系的方式。他们并不能取代任何老的联系方式。

1. C 2. B 3. A 4. D

Section Four Writing

Activity Schedule	
Monday, April 18	
4.00 p.m.	Arrive in Beijing by Flt.CZ8911, to be met at the airport by Mr. Yang Guang, President of Asia Trading Co.
4.15 p.m.	Leave for Great Wall Hotel by Car
7.30 a.m.	Dinner Party given by President Yang Guang
Tuesday, April 19	
9:30 a.m.	Discussion at Asia Trading Co. Building
2:00 p.m.	Group discussion
7:00 p.m.	Watching Performances in the Lao She Teahouse
Wednesday, April 20	
9:00 a.m.	Discussion
12:00 noon	Sign the Letter of Intent
1:30 p.m.	Peking Duck Dinner
3:30 p.m.	Visit the Summer Palace
6:00 p.m.	Departure for Shanghai by Air

Section Five Grammar

I. 1. going 2. coming 3. running 4. leaving 5. regretting 6. living

II. 1. driven 2. broken 3. hurt 4. weighed 5. regretted 6. lived

III. 1. pleasing 2. retired 3. given 4. interesting 5. written 6. rising
　　7. spoken 8. confusing 9. exhausting 10. satisfied

IV. 1. C 2. D 3. A 4. C 5. C 6. B 7. D 8. A

9. D   10. A   11. C   12. D   13. A   14. B   15. D

Unit Four Computer Network

Section One   Warming Up
1. protocol   2. browser   3. website   4. domain name

Section Two   Real World
Find Information
Task I:
1. She works in the Customer Service.
2. He makes an attempt at connecting to the Internet through ADSL, but he can't access the Internet. He wants to solve the problem.
3. He should first check whether his ADSL modem connects to the telephone line in the right way.
4. His ISP.
5. He should let the system automatically obtain its IP and DNS address.

Task II:1.T 2. F 3. F 4. T 5. T

Words Building
Task I:1.B 2. A 3.B 4.D 5. C
Task II:1.attempting 2.is activated 3.to configure 4.obtains 5.addressing
Task III:1.B 2. F 3.D 4. J 5.A 6.G 7. I 8. C 9.E 10.H

Cheer up Your Ears
TaskI: 1.make an attempt 2.access 3.modem 4.been activated 5.username and password 6.correct 7.configure 8.adapter 9.automatically 10.get
Task II:1.Network 2.convenience 3.deal with 4.you feel tired 5.virus websites
Task III: 1. A 2. C 3. B 4. C 5. D

Table Talk
Task I: 1.What can I do for you?   2. if you have all the necessary equipment   3.network adapters 4. modem 5. the wireless network

Section Three Brighten Your Eyes

宽带概述

　　宽带是指那些快速的、持续的互联网连接。可以通过电缆、卫星、移动电话、光纤或者是ADSL 线路传输（宽带）信号。

　　拨号上网和宽带的区别就像是乡间小路和高速公路的区别。宽带用分开的频带取代单一频带来进行数据上传、下载和语音传输。ADSL 信号来源于当地的电话交换系统，该系统通过传统的铜质线缆传输。您的互联网服务提供商会为您测试和激活线路。一些有线电视公司通过它们现有的线路提供有线互联网连接。光纤是细小的导线，发射光脉冲。因为光是宇宙中速度最快的东西，它们比 ADSL 电缆快得多。这意味着比起最快的 ADSL 连接上网速度，您可以以其高达五倍的速度上网。移动连接是最新的宽带连接方式。一个小的 USB 设备或者是数据卡能让您在有手机信号的地方就能访问互联网。

Find Information
Task I:
1. Broadband is the name given to any fast, permanent Internet connection.
2. It can be delivered by cable, satellite, mobile phone and ADSL.

3. Your ISP will test and activate the line for you.
4. Yes, they are.
5. Mobile is the newest form of the broadband.

Task II: 1. F 2. F 3. F 4. T 5. T

Words Building

Task I:

1. Internet connection	2. mobile phone signal
3. dial-up	4. country lane
5. local phone exchange	6. cable TV
7. USB device	8. data card

Task III: 1. B 2. C 3. C 4. D 5. D

Task IV: 1. permanently 2. difference 3. dialed/ dialled 4. uploading 5. locally

Cheer up Your Ears

Task I:

1. fast, permanent	2. cable, satellite, mobile phone
3. a country lane, a motorway	4. uploading, downloading data
5. via	6. existing wiring
7. activate the line	8. newest
9. USB device	10. phone signal

Task II：1. C 2. D 3. C 4. A 5. B

Listening

Task I 1. D 2. B 3. D 4. A 5. C

Task II.1. general manager 2. business 3. sales people 4. secretary 5. promoted

Extensive Reading

　　我们都在忙着谈论和使用互联网，但我们有多少人知道互联网的历史？当许多人知道互联网是建立在20世纪60年代的时候，都感到很惊讶。在那个时候，电脑又大又贵。计算机网络运行并不顺畅。如果网络上的一台电脑崩溃了，那么整个网络就崩溃了。这样就必须建立一个新的网络系统。好的计算机网络应该是由很多台不同的电脑组成的。如果网络的一部分不运转了，信息可以通过另一部分发送。在这种方式下，计算机网络系统将持续运行。

　　起初，只有政府使用互联网，但在70年代初，大学、医院和银行也被允许使用它。然而，电脑仍然非常昂贵，互联网是难以使用。直至上世纪90年代初，电脑变得更便宜，更容易使用。科学家们还开发了软件，使得浏览互联网变得更加方便。

　　如今，上网很容易。据说，每一天都有数以百万计的人在使用互联网。发送电子邮件在学生中越来越流行。互联网现在已经成为人们生活中最重要的部分之一。

1. C 2. B 3. A 4. C 5.C

Section Four Writing

鲨鱼油
主要原料：鲨鱼油
功效成分：每1 000毫克中含DHA 250-300毫克，EPA 60-90毫克
保健作用：改善记忆，调节血脂
剂量与用法：口服，初服者，每日2次，每次2粒；长期服用者，每日1次，每次1-2粒。
贮存：置于阴凉干燥处
有效期：二年

参考答案

Section Five　Grammar

① - ⑤ 1. which / that　2. where / in which　3. which / that / ×
　　　 4. which / that / ×　5. which

I. 1. which / that　2. who / whom / that / ×　3. which / that　4. which / that / ×
　 5. who / that　6. where / in which　7. who / that　8. who / whom / that / ×
　 9. whom　10. who / that

II. 1. D　2. B　3. C　4. C　5. A　6. B　7. A　8. A　9. D　10. D

III. 1. whose　2. who / that　3. that / which　4. who / that　5. where
　　 6. why　7. who / that　8. that / which　9. when　10. where

Unit Five　Network Security

Section One　Warming Up

1.firewall　2. antivirus　3. spam　4. virus

Section Two　Real World

Find Information

Task I:

11. She works in the Technical support department.

12. He wants to know how to set up a firewall for his Windows system.

13. Yes, he can.

14. He can install antivirus software on the computer.

15. He can buy some commercial firewall products and some other network security software.

Task II: 1.F　2. F　3. T　4. F　5. T

Words Building

Task I:

1. B　2. B　3.A　4.B　5. D

Task II: 1.connect　2.advanced　3.able　4.secure　5.commercially

Task III: 1.G　2. F　3.H　4. B　5. I　6. C　7. J　8. A　9. E　10. D

Cheer up Your Ears

Task I: 1.Technical　2.firewall　3.enable　4.click　5.advanced
6.disabled　7.antivirus　8.security　9.commercial　10.see

Task II: 1.access　2. instructions　3. setting　4. browsers　5. password

Task III: 1. C　2. D　3. A　4. B　5. C

Table Talk

1. The system crashed　　　　　　　2. illegal websites.

3. what's wrong with my PC　　　　 4. antivirus software

5. what about the data I stored in the computer?

Section Three　Brighten Your Eyes

网络安全管理

　　计算机安全是预防和检测未经授权而使用您的计算机的过程。预防措施有助于阻止未经授权的用户（也被称为"入侵者"）访问您的计算机系统的任何部分。一个有效的网络安全策略需要识别威胁，然后选择最有效的工具来打击它们。

　　不同情况下网络安全管理不尽相同。家庭或办公室的小型网络只需要基本的安全防护，而

10

参考答案

大型商业网络则需要更高维护性和更先进的软件来阻止黑客和垃圾邮件的恶意袭击。

对于小型家庭网络，每台连接到互联网的计算机都需要防火墙的保护，无线网络则需要双层或3层防火墙。您可以在每台计算机上安装软件防火墙，也可以在整个网络中安装硬件防火墙。

Find Information

Task I: 1. No, it is different.
2. A small home or an office would only require basic security.
3. Large businesses will require high maintenance and advanced software and hardware.
4. Yes, it is.
5. We can either install software firewalls on each computer, or install a hardware firewall on your entire network.

Task II: 1. F 2. F 3. F 4. T 5. T

Words Building

Task I: 1. Security Management   2. large businesses network
3. basic security   4. high maintenance
5. advanced software and hardware 6. malicious attack
7. firewall   8. wireless network

Task III: 1. B 2. A 3. C 4. D 5. A

Task IV: 1. manager 2. difference 3. requirements 4. hacker 5. protection

Cheer up Your Ears

Task I: 1. Security Management   2. basic security
3. advanced software and hardware 4. hacking and spamming
5. connected to the Internet   6. double pay
7. a triple room   8. wireless network
9. install software firewalls   10. install a hardware firewall

Task II: 1. C 2. D 3. B 4. A 5. A

Listening

Task I. 1. D 2. A 3. C 4. D 5. A

Task II. 1. introduce 2. sell 3. North America 4. customers 5. look forward to

Extensive Reading

电子邮件使用安全须知

以下是一些简单的方法，用以保护您在家使用电子邮件时的安全：

- 不要在邮件中发送机密信息，不要对那些您不信任的未知邮件或网络地址回复带有个人信息的邮件。
- 不要假定邮件中的建议或指令有好的意图。
- 对待从您不认识的人或组织发送过来的邮件要非常小心。不要打开任何邮件中的附件，直到它通过最新的安全软件如诺顿互联网安全软件的扫描。
- 安装垃圾邮件过滤器，这样您就不会收到包含有害内容的无用邮件。

Task I. filter; intention; instruction; confidential

Task II. 1. d 2. c 3. a 4. B

Section Four Writing

I. 1. D 2. G 3. A 4. J 5. E 6. C 7. I 8. F 9. B 10. H

II. 1. 旅客登记处 2. 此路不通 3. 停车场 4. 失物招领处 5. 油漆未干

6. 损坏赔偿 7. 入口 8. 禁止倒垃圾 9. 不准带出室外 10. 免费上网

Section Five Grammar

I. 1. (that) 2. that 3. (that) 4. (that) 5. (that) 6. that 7. which
8. which 9. why/ for which, (that) 10. whose 11. As
12. where/ in which,(that) 13. who 14. whom 15. whom

II. 1. 去掉 it 2. girl 后加 who / that 3. who 改成 whom 4. that 改成 which 5. have 改成 has
    6. which 改成 that or 7. which 改成 that or 8. that 改成 which 9. Who 改成 Those who
    10. whom 改成 that or

III. 1. B 2. A 3. C 4. D 5. D 6. B 7. D 8. D 9. C 10. B 11. C 12. D 13. B 14. C 15. B

Unit Six Virus

Section One Warming Vp
1. F 2. G 3. I 4. J 5. H 6. A 7. B 8. E 9. C 10. D

Find Information
Task I.
1. Her computer crashed.
2. It might have been infected with a virus.
3. No, it shouldn't.
4. Because E-mails are not system files.
5. To try some anti-virus software like Kaspersky.
Task II. 1. T 2. F 3. F 4. T 5. T

Words Building
Task I. 1. A 2. B 3. A 4. B 5. D
Task II. 1. attachment 2. infection 3. container 4. exact 5. essentially
Task III. 1. B 2. G 3. A 4. H 5. F 6. C 7. I 8. J 9. D 10. E

Cheer up Your Ears
Task I. 1. crashed 2. infected 3. spread 4. essential 5. at once
        6. executable 7. exactly 8. system 9. contain 10. anti-virus
Task II. 1. in a mess 2. sorry to say 3. latest type 4. check 5. at your service
Task III. 1. C 2. B 3. D 4. A 5. D

Table Talk
1. take a look at 2. give me a second 3. infected files
4. have no idea 5. make sure

Section Three Brighten Your Eyes

病毒对计算机系统的影响

　　正如我们现在所知，计算机病毒起源于 1986 年第一个个人电脑病毒 Brain 的产生。两兄弟写了这个病毒（Basid 与 Farrop Alvi，二人在巴基斯坦的拉合尔经营一家小软件公司），随后开始了今天仍在继续的病毒和杀毒程序之间的竞争。病毒可以被重新编程，造成多种危害，包括以下内容。

　　自我复制到磁盘的其他程序或区域。

　　快速频繁的复制，侵占受感染的系统磁盘和内存，致使系统无法工作。

　　修改，损坏或摧毁选定的文件。

清除整个磁盘的内容。
在指定时间内潜伏并在特定的条件下激活。
打开通向被感染系统的后门，让别人通过网络或 Internet 连接来访问甚至控制该系统。
窃取系统或用户的重要信息并发送到指定位置。

Find Information
Task I. 1. In 1986. 2. Brain. 3. Eight. 4. No. 5. By causing programs to behave oddly.
Task II. 1. T 2. T 3. F 4. F 5. T

Words Building
Task I.
1. personal computer   2. anti-virus program   3. disk and memory
4. selected file   5. erase the content   6. lie dormant
7. access the system   8. steal information
Task III. 1. A 2. B 3. A 4. C 5. C
Task IV. 1. harmful 2. frequent 3. entirely 4. connect 5. Originator

Cheer up Your Ears
Task I. 1. harm 2. worried about 3. turned on 4. frequently 5. active
         6. copies 7. perform 8. messages 9. originated 10. passwords
Task II. 1. B 2. B 3. C 4. D 5. A

Listening
Task I. 1. A 2. D 3. B 4. B 5. C
Task II. 1. order 2. clearly 3. another 4. as much 5. mouth

Extensive Reading
　　病毒常常会对受感染的主机进行某种有害的活动，这有一些保护您个人电脑的小窍门。
- 始终更新您的操作系统和应用程序。
- 安装杀毒和反间谍软件，在系统启动时自动扫描病毒。
- 每天更新病毒定义，确保您的系统免受最新的威胁。
- 不要从互联网下载任何文件，除非您确定不会传送病毒。
- 不要使用任何已在另一台计算机使用的存储介质，除非您确定其他电脑无病毒，不会把病毒传染给您的系统。
- 不要打开您不知来源的电子邮件附件；不要打开任何 .exe 附件。运行或安装下载文件前件应该用杀毒程序扫描。

1. up-to-date   2. Anti-spyware   3. daily   4. download
5. storage media   6. free   7. attachments   8. scanned

Section Four Writing
To: Mr. Wang Jin, President of Shanghai University
From: Zhang Li, Dean of studies
Date: May 6, 2006
Subject: <u>Buying computers and DVs</u>
　　Upon the request of the Equipment Division of the University, we have inspected the lab of the Mathematics Department and found its present equipments unsatisfactory to students, particularly to postgraduates. <u>In order to improve the effect of experiments, it is hereby recommended that ten computers</u> and 2 DVs be bought and issued to the laboratory.

参考答案

Section Five Grammar

I. 1. is   2. were   3. would have given   4. would feel   5. would have lost
   6. had   7. had been   8. loved   9. Were   10. hadn't told
II. 1. D   2. D   3. C   4. C   5. C   6. C   7. D   8. B   9. D   10. B
III. 1. have——had   2. came——had come   3. will come——came
    4. haven't——hadn't   5. Were it rain——Were it to rain/ Should it rain
IV. 1. could fly   2. But for   3. were, would go   4. were
    5. as though/ as if   6. were   7. Were I
    8. If it should rain/ Were it to rain   9. Had I known   10. Without
V. 1. could   2. lived   3. could   4. were   5. will come

Unit Seven Website

Section Two Real World
Find Information
Task I.
6. He wants to create a website for his company.
7. The dot com.
8. No, it isn't.
9. You may be forced to display advertising and free services are usually less reliable and slower.
10. Offering valuable and attractive information.
Task II. 1. T   2. F   3. F   4. T   5. F

Words Building
Task I. 1. B   2. B   3. C   4. B   5. B
Task II. 1. recommendation   2. register   3. geography   4. creative   5. value
Task III. 1. D   2. G   3. I   4. J   5. A   6. F   7. B   8. C   9. H   10. E

Cheer up Your Ears
Task I. 1. domain   2. inexpensive   3. recommended   4. paid   5. forced
        6. reliable   7. high   8. more than   9. little   10. valuable
Task II. 1. website   2. as important as   3. description   4. difficult   5. a lot of fun
Task III. 1. B   2. D   3. A   4. A   5. C

Table Talk
1. creating a website   2. in a week   3. communicate with others
4. contact   5. Thanks for

Section Three Brighten Your Eyes

网站设计技巧

在设计您的网站时，应该遵循一些基本规则。这些技巧将使您的网页访问者更容易在网上冲浪。

技巧1. 访客至上的原则。

在您动笔之前，您应该知道它们的需求。看它们寻找什么并给它们提供什么。通过增加它们寻找的信息使您的站点有吸引力。

技巧2. 使您的网站容易阅读。

网站可能由于文本选择了不合适的字体和颜色而变的糟糕。

14

技巧 3．您的网站应该容易操作。

您可以使用图表图象，譬如按钮和制表符，或文本链接。无论使用何种链接，保证它们清楚而准确地被标记并且访客容易识别。确保所有的页面都有回到主页的链接。

技巧 4．您的的网站应该容易找到。

在网上推广自己网站最容易的方式是通过搜索引擎、目录、有奖站点、电子邮件和其它网站链接。网站上应有您的联系信息以防有人需要联系您。

技巧 5．保持您的网页布局和设计一致和连贯。

技巧 6．您的网站应该快速下载。

避免使用过多图表、动画，由于这些文件会占用服务器的带宽。您的网页应该在 10 秒内打开，否则访客会有沉闷感和挫折感。

Find Information

Task I.
1. Website design tips.
2. By adding information that they are looking for.
3. We should make sure they are labeled clearly and accurately and familiar to your visitor.
4. Through search engines, directories, award sites, emails as well as links from other websites.
5. Because they eat up bandwidth with the server and the page can't display quickly.

Task II. 1. F 2. F 3. T 4. T 5. F

Words Building

Task I. 1. website design 2. font and color choice 3. text link 4. main page
        5. search engine 6. award site 7. contact information 8. web page layout

Task III. 1. D 2. B 3. C 4. A 5. B

Task IV. 1. attract 2. navigation 3. accurate 4. consistently 5. frustrate

Cheer up Your Ears

Task I. 1. basic rules 2. attractive 3. worse 4. navigate 5. accurately
        6. main page 7. in case 8. consistent 9. display 10. eat up

Task II. 1. D 2. B 3. B 4. B 5. C

Listening

Task I. 1. B 2. A 3. C 4. A 5. C

Task II. 1. latest 2. hardware 3. old people 4. eyesight 5. give out

Extensive Reading

当您设计新网站时，您会被一些致命错误的网页设计想法所诱惑。以下是不惜一切代价要避免的最常见的 5 个错误。

1. 太多的图表

太多的图表可以使您的网站加载速度太慢。游客会变得不耐烦，经常点击一次您的网站一永远不回来。

2. 计数器

访客计数器或点击计数器不应该显示在您的网站上，除非您有巨大的游客量。唯一显示计数器的好处是您有数以百万计的游客，想显示您网站的人气或想吸引大量广告商。

3. 横幅广告

限制您的网页页面的横幅广告，除非是必需的。为什么？因为横幅广告会减慢加载速度，让网上冲浪者关闭网页。最重要的是，"横幅广告"仅仅是"广告"的代名词，人们不愿意点击。

15

参考答案

4. 网站散乱
　　规划您的网站来引导游客是很重要的，不管您是引导他们去买东西，或是点击进入您的网站的另一个地方。
5. 泛化
　　在互联网上销售的最有效的方法是使您的网站个性来确定您的目标受众。
1. slow　2. impatient　3. tremendous　4. popularity
5. bare necessities　6. clicking　7. lead visitors　8. target audience

Section Four　Writing

> Griffith University
> Faculty of Education and Arts
> Gold Coast Campus Parklands Drive Southport
> Telephone: 076948803　　Fax: 076948534
> E-mail: petergreen@eda.gu.edu.au
>
> To: <u>Dingxiao</u>
> （北京大学英语系）<u>English Department</u>
> 　　　　　　　　<u>Beijing University</u>
> Fax: <u>0411-76213453</u>
> From: <u>Peter Green</u>
> Date: <u>Oct. 10, 2015</u>
> Pages: 1
> Re: <u>Your Arrival</u>
> 　（亲爱的丁晓：）<u>Dear Dingxiao,</u>
> 　（谢谢您2015年9月12日的来信。我理解您的问题。您可以12月来此。请随时告知。）<u>Thank you for the letter of Sept. 12, 2015. I understand your problem. A December arrival is fine. Please keep me informed.</u>
> 　　　　　　　　　　（祝好。）<u>Best wishes,</u>
> 　　　　　　　　　　（彼特·格林）<u>Peter Green</u>

Section Five　Grammar

I. 1. (should) give　2. (should) be put　3. (should) finish
　4. (should) be solved　5. (should) be allowed　6. left　7. took
　8. (should) make　9. (should) be finished　10. (should) miss
II. 1. B　2. D　3. C　4. D　5. C　6. C　7. C　8. C　9. B　10. B
III. 1. 把 will 改成 should 或去掉 will.　2. worked——(should) work
　3. does——did　　　　　　　　　4. take——be taken
　5. cleans——(should) clean　　　　6. must be——must have been
IV. 1. were　2. decided　3. (should) try
　4. (should) stay　5. (should) go　6. (should) be reserved

Unit Eight　Search Engine

Section Two　Real World
Find Information
Task I. 1. A cell phone.

2. He advises Candy to use a search engine and have a look at the evaluation and recommendation first.

3. In 1990.　4. Yes.　5. They are students.

Task II. 1. T　2. T　3. F　4. F　5. T

Words Building

Task I. 1. B　2. D　3. A　4. C　5. B

Task II. 1. evaluate　2. organize　3. complicate.　4. technological　5. incredibly

Task III. 1. G　2. I　3. J　4. E　5. A　6. F　7. H　8. B　9. C　10. D

Cheer up Your Ears

Task I. 1. variety　2. have a look　3. search for　4. came out

　　　　5. take time　6. rank　7. make money　8. Provide

Task II. 1. millions of　2. search engine　3. interested　4. less than

　　　　5. at the same time

Task III. 1. D　2. A　3. A　4. B　5. A

Table Talk

1. think of　2. In my opinion　3. at least　4. personal privacy　5. What's more

Section Three　Brighten Your Eyes

搜索引擎如何运行

　　网页搜索引擎运行时，存储从网页 HTML 检索到的信息。这些网页由网络爬虫检索（有时也称为蜘蛛）。

　　然后搜索引擎分析每一页的内容来确定应该如何索引（如可以从标题、网页内容，或叫作元标记的特殊领域提取单词）。关于网页的数据都存储在索引数据库中以便以后查询。用户的查询可以是一个词。检索有助于尽快找到与查询相关的信息。

　　当用户在搜索引擎输入查询时，引擎检查其检索。然后根据其标准提供了最佳匹配的网页清单，通常显示文件的标题，有时候是一部分正文。

　　一个搜索引擎的有效性取决于它所返回的结果集的相关性。尽管可能有数以百万计的网页包含某个特定的词或短语，有些网页可能更相关，更流行，更权威。大多数搜索引擎采用某些方法进行排序来提供"最好"的结果。

Find Information

Task I. 1. How search engines work.　　2. By a Web crawler.

　　　　3. In an index database.　　　　4. A listing of best-matching web pages.

　　　　5. The relevance of the result set it gives back.

Task II. 1. T　2. F　3. F　4. F　5. T

Words Building

Task I. 1. search engine　2. web crawler　3. meta tag　4. enter a query

　　　　5. a listing of best-matching web pages　6. the document's title

　　　　7. result set　8. rank the results

Task III. 1. C　2. C　3. A　4. B　5. C

Task IV. 1. retrieval　2. determination　3. relate　4. crawl　5. relevant

Cheer up Your Ears

Task I. 1. storing　2. crawler　3. analyses　4. database　5. relating to

　　　　6. query　7. best-matching　8. depends on　9. relevant　10. rank

Task II. 1. B　2. C　3. B　4. D　5. C

17

参考答案

Listening
Task I. 1. A 2. D 3. B 4. C 5. D
Scripts:
Task II. 1. day 2. as long 3. boring 4. far away 5. discuss
Extensive Reading

人肉搜索引擎是一个中文词语，利用博客和论坛等网络媒体进行调查研究。该系统基于大量的人力合作。它一般被视为用于向公众暴露个人隐私。然而，最近的分析表明它也可以用于其他一些原因，包括揭露政府腐败，确定肇事司机，揭露科学造假，以及更多的"娱乐"事件例如识别照片上看到的人。

由于互联网的便捷和高效性，"人肉搜索"经常被用来获取一些通过其他常规手段（如图书馆或网络搜索引擎）很难或不可能得到的信息。一旦可用，这样的信息，可迅速分布到数百个网站，使其成为一个非常强大的大众媒介。由于获得信息要依靠个人知识或非正式的（有时是非法的）访问，搜索的可靠性和准确性往往有所不同。

1. massive human collaboration   2. individual privacy
3. government corruption   4. hit and run drivers
5. scientific fraud   6. difficult or impossible
7. personal knowledge   8. unofficial access

Section Four Writing

> Speech of New Year
>
> Ladies and Gentlemen,
>   Happy New Year!
>   This is the first new year since our company's opening here. As I look back over the past year, I am grateful to you all for your efforts which have made it possible for the company to achieve unexpectedly high performance. I thank you again.
>   I hope that this year will be an even greater year for our company and for all of us, and finally I wish all of you present here good health and happiness.

Section Five Grammar
1. C 2. A 3. A 4. D 5. D 6. C 7. B 8. D 9. A 10. D
11. C 12. A 13. B 14. C 15. A 16. D 17. D 18. D 19. D 20. B

Unit Nine Network and System Administrator

Section One Warming Up
1. C 2. F 3. E 4. J
Section Two Real World
Find Information
Task I.
1. He's required to set up the computer system and the network.
2. Yes, it is.
3. Its speed is much slower and can be easily affected by the surrounding and has some security risks.
4. The PDF reader
5. Yes.

Task II. 1. F 2. T 3. F 4. T 5. T
Words Building
Task I. 1. B 2. A 3. D 4. C 5. C
Task II. 1. wire 2. surrounded 3. flexibly 4. accessible 5. installing
Task III. 1. B 2. E 3. H 4. D 5. C 6. J 7. I 8. A 9. G 10. F
Cheer up Your Ears
Task I. 1. set up 2. meet 3. computer systems 4. password 5. access
  6. surroundings 7. install 8. desktop 9. manage 10. feel free
Task II. 1. work 2. followed 3. turning it off 4. before 10 o'clock 5. connect
Task III. 1. D 2. B 3. A 4. B 5. C
Table Talk
1. take care of 2. what's the matter 3. network connection
4. write it down 5. a little bit
Section Three Brighten Your Eyes

网络和系统管理员

在许多环境中都可以找到网络和系统管理员，包括工业、政府、制造业、教育相关的环境。他们通常在舒服的办公室或机房工作。一般每周工作40小时；然而许多管理者往往在晚上或周末因任务面"随叫随到"。由于可能会有意想不到的计算机或网络的技术问题，加班是常有的事。担任客户咨询顾问的管理人员可以不呆在自己的办公室，可能每次需要在客户的办公室工作几周甚至几个月。现在管理人员往往能够提供远程技术支持，而不必前往顾客的工作地点。

这项工作往往很有趣，并且薪水一般来说也很优厚。对于那些长期在键盘上打字的人来说，缺点是会造成背部不适，眼疲劳，手或腕部疾病。工作中经常遇到的最后期限以及注重准确性和精确性可能是引起压力的原因。另外，需要不断地与他人进行交流，在感到满意的同时，有时也会令人沮丧，当计算机系统发生故障时，情况尤为如此。

Find Information
Task I.
1. Industry, government, manufacturing, and education.
2. Forty hours.
3. Yes. Because unexpected computer or network technical problems arise.
4. Because they can provide technical support from remote locations.
5. Back discomfort, eyestrain, hand/wrist problems.
Task II.
1. F 2. F 3. T 4. T 5. T
Words Building
Task I. 1. network and system administrator 2. computer laboratory
  3. on call 4. technical problem 5. technical support
  6. remote location 7. type for long periods 8. back discomfort
Task III. 1. B 2. B 3. D 4. C 5. C
Task IV. 1. environmental 2. consult 3. stressed 4. comfort 5. constant
Cheer up Your Ears
Task I. 1. situation 2. must 3. environments 4. possible 5. consultants
  6. interact 7. duties 8. extended 9. management 10. administrator
Task II. 1. D 2. A 3. A 4. A 5. C

参考答案

Listening

Task I. 1. D 2. B 3. B 4. C 5. D

Task II. 1. rich   2. 1998   3. growing   4. less expensive   5. enjoy

Extensive Reading

　　网络和系统管理人员的首要职能是安装、支持和维护计算机系统的组件，其中包括服务器和网络元件。此外，管理人员通常要负责了解最新技术，确定如何配置设备，以便更好地为机构的短期和长期目标服务。其职责可能包括以下几个方面：

- 维护网络的硬件和软件
- 确定系统硬件和软件的所需条件
- 确保所有系统用户可连续使用网络
- 规划和执行网络安全措施
- 为改善系统和网络配置提出整改建议
- 为修复工作保存库房备件
- 实施技术更新、安装补丁和变更配置
- 进行资料日常备份
- 协调计算机网络连接和使用

1. Maintenance   2. requirements   3. availability   4. implementation
5. Recommendation   6. repairs   7. Implementation   8. Performance

Section Four   Writing

```
                        Diploma
Xu Tingting, female, born on September 12, 1986, native of Shanghai, was an
undergraduate student who was studying Foreign Trade of Pujiang University
during September, 2008 to July, 2012. She has completed all the prescribed four-
year-undergraduate courses, has passed all the examinations and is expected to be a
graduate of Pujiang University.
                                              Signature
 _____
 No. 1234567890                       President of Pujiang University
                                       Issued on July 22, 2012
```

Section Five   Grammar

1.A 2.A 3.C 4.B 5.C 6.B 7.A 8.A 9.A 10.B
11. B 12.D 13.C 14.B 15.D 16.B 17.A 18.D 19.C 20.D

Unit Ten IT Jobs

Section One Warming Up

1. A. Web Designer（网站设计师） B. Web Editor（网站编辑）
　 C. Website Promotion（网站推广） D. Network Administrator（网络管理员）
　 E. Data Analyst（数据分析） F. Graphic Designer（面设计师）
　 G. Programmer（程序员） H. Software Developers（软件开发）
　 I. Software Testing Engineer（软件测试工程师）

2. 1). ID Card 2). Resume 3). Certificate 4). Diploma

参考答案

Section Two Real World
Find information
Task I. 1. China Daily.
　　　　2. Computer Programmer.
　　　　3. College education on computer science with some work experience and good English.
　　　　4. $2,000 a month.
　　　　5. Fill in the application form on the company's website.
Task II. 1. F 2. T 3. T 4. F 5. F
Words Building
Task I. 1. B 2. D 3. B 4. B 5. B
Task II. 1. qualified 2. temporary 3. sick 4. application 5. basically
Cheer Your Ears
Task I. 1. programmer 2. opening 3. qualifications 4. applicants 5. requirement 6. permanent
　　　　7. insurance 8. apply 9. application 10. Information
Task II. 1. interview 2. apply for 3. computer program 4. suitable 5. excellent
　　　　6. sort of 7. in charge of 8. laid off 9. advance 10. decision
Task III. 1. D 2. B 3. A 4. C 5. A
Table Talk
1. software engineering 2. work experience 3. intern 4. spoken English
5. opportunity to advance
Section Three Brighten Your Eyes

大有前景的工作

　　拥有计算机技能及培训经历的求职者们极有可能大量的工作机会。让我们熟悉一些信息工程相关工作吧。
- 软件工程师
- 软件工程师为网络、数据系统、网站，更多的是为如智能手机或平板电脑等移动设备设计应用程序，以苹果公司的 iPad 为例创造程序。在苹果公司吹嘘"定有一个应用适合您"的时候，就意味着消费者们有无数的应用可以使用。这也意味着单单为某一公司设计这些应用程序的人们供不应求。
- 运筹分析师
　　这一头衔对于那些不熟悉的人来说听起来有点儿神秘，但基本上这些人们用数学、计算机建模、编程和其方式的定量分析方法解决问题。运筹学方法原本是为军方所用但现在为更多的公私机构使用。随着各组织机构寻求高效，运筹分析师将有极佳的工作前景。
- 系统分析员
　　不论系统分析员设计从零开始设一个计算机系统还是修改一个已存的系统，他们必须顾全大局。这份工作需要对一组织机构如何运作有透彻的理解，才能选择出更好服务与这一机构的硬件及软件。

Find Information
Task I.
1. Software engineers create applications for networks, data system, the Web and mobile devices.
2. They use math, computer modeling, programming and other types of quantitative analysis to solve problems.
3. Yes, they do.

参考答案

4. They design a computer system from scratch or tweak an existing one.
5. A thorough understanding of how an organization functions.
Task II. 1. F 2. T 3. T 4. T 5. F

Words Building
Task I. 1. mobile devices 2. smart phones 3. computer modeling
       4. quantitative analysis 5. operational research
Task III. 1. A 2. B 3. A 4. D 5. B
Task IV. 1. skillful 2. quantity 3. mystery 4. analyze 5. efficiency

Cheer up Your Ears
Task I. 1. opportunities 2. applications; devices 3. design; demand
       4. uninitiated; analysis 5. operational; range 6. prospects; efficient
       7. scratch 8. functions
Task II. 1. D 2. C 3. B 4. C 5. C

Listening
Task I. 1. B 2. C 3. A 4. D 5. D
Task II. 1. feet 2. warm 3. next to 4. swimming 5. avoid

Extensive Reading

10个面试问题

做好准备有效地回答面试官在面试时通常会问的问题很重要。因为这些问题如此寻常，招聘经理就会期望您能毫不犹豫地顺利地回答这些问题。让我们复习这十个您在面试时常见被问到的问题及回答。

1. 您的强项是什么？
    当被问及强项时，提到能满足工作要求的特长，以使您区别于其他面试者。
2. 您的弱项是什么？
    努力把回答往积极的方向引导，如您的技能或特长等。
3. 为何离职？
    实事求是，坦诚，把重点放在今后工作上，特别是在您的离职不是好的状况下。
4. 谈谈您自己。
    以不直接涉及工作的个人兴趣作为开始。
5. 为何想从事这份工作？
    仔细描述您符合工作要求的特质，并提到公司和职业吸引您的地方。
6. 我们为什么要雇用您？
    用简洁的回答来推销自己，并解释您能为公司提供什么和为什么您应该得到这份工作。
7. 如何处理压力？
    最好以您之前工作时如何处理压力作为回答的方式。
8. 说说您遇到的困难，并且如何克服它的。
    因为这一问题也是关于压力的，分享您在艰难环境下的所作所为。
9. 您如何定义成功？
    这样的问题是雇主想看看您作为雇员的工作态度、目标和整体素质。考虑一下您的公司和职位，把回答基于这两点并且描述一下您的个人价值观和目标。
10. 您的未来目标是什么？
    把回答聚焦于您面试的工作和公司。

1. F 2. T 3. T 4. F 5. T 6. F 7. T 8. T

Section Four Writing

1. The Foreign Teacher shall <u>be entitled</u> to the following <u>additional benefits</u>. This School provides these benefits <u>at no cost</u> the Foreign Teacher. No <u>compensation</u> given to those who do not choose to participate or avail themselves of these benefits.
2. The Foreign Teacher shall <u>submit a written application for sickness</u> if unable to work <u>prior to</u> the concerned class time. This School may require the Foreign Teacher to see a doctor prior to <u>giving permission</u>. Teachers will be allowed <u>2 sick days leave with pay per month</u>. The Foreign Teacher may not <u>accumulate</u> unused sick days.

Section Five Grammar

1. C 2. D 3. A 4. A 5. B 6. B 7. B 8. A 9. C 10. B
11. A 12. C 13. A 14. B 15. C 16. C 17. D 18. C 19. D 20. B

Who is the girl (that) we saw just now?
我们刚刚看见的那个女孩是谁？

⑧在 There be/ live 句型中，先行词指物时。例如：
There is a book on the desk (that) I'm interested in.
有一本我感兴趣的书在桌子上。

⑨当先行词前有 such 或 the same 修饰时，原则是用 as 来引导，但当先行词与关系代词指同一物时，则用 that. 如：
This is the same watch that I lost three weeks ago.
这就是我三周前丢的那块表。

3. 只用 who 的情况

①先行词为 those 且指人时。例如：
Those who break the law will be punished. 违法的那些人会受到惩罚。

②先行词为表人的不定代词 one, anyone, nobody, anybody 等时。例如：
Anyone who breaks the law will be punished.
任何违法的人都会受到惩罚。

③ There be 句型中的先行词指人时。例如：
There's a man outside who wants to see you. 外面有人想见你。

④在分隔定语从句中，若先行词表人时，为了避免混淆，用 who。如：
He's the only person in our office who has been invited to the dinner party.
他是我们办公室里唯一被邀请参加晚宴的人。

4. whose 引导的定语从句能与 of which 或 of whom 结构互换。例如：
I saw some houses whose windows all face south.
我看见一些房子的窗户朝南。
(此句中的 whose windows 可换成 the windows of which 或 of which the windows)

5. as 引导的定语从句的几种情况

①先行词前有 such 修饰时。例如：
There's no such place as you dream of in the world.
世界上没有像你梦中一样的地方。

②先行词前有 the same 修饰时。例如：
This is the same watch as I lost three days ago.
这块表和我三天前丢的那块相似。
但当关系词和先行词指同一事物时，应用 that 来引导。例如：
　This is the same watch that I lost three days ago.
（这就是我三天前丢的那块表）

③当译为"正如"时。应用 as 来引导非限定性定语从句。例如：
As we all know, Taiwan is part of China.
众所周知，台湾是中国的一部分。

6. 介词短语不能拆开，即介词不能提前。例如：
Is this the book for which you are looking?（此句错误）
应改为：Is this the book you are looking for?（因为词组不能拆开）

这是你正在找的那本书吗?

7. 定语从句特例:

①当 way 作先行词时,有 3 种方法,即 in which /that / ×（不填）。例如:I don't like the way in which/that/ × he speaks.

我不喜欢他说话的那种方式。

② Tom is one of the students who are good at English.

汤姆是擅长英语的学生之一。

（当先行词前有 one of 修饰时,定语从句的谓语动词应用复数）

但当 one of 前面有 the 或 the only 的时候,则用单数。例如:

Tom is the only one of the students who is good at English.

汤姆是唯一擅长英语的学生。

Exercises

I. Fill in each blank with proper relatives.

1. Is there anything else _____ you need?

2. They talked happily about the men and books _____ interested them.

3. I'll tell you all _____ he told me last month.

4. This is the most interesting book _____ I have ever read.

5. The last place _____ we visited was the Great Wall.

6. Tom is the very man _____ is fit for the job.

7. Football, _____ is a very interesting game, is played all over the world.

8. His glasses, without _____ he was like a blind man, were missing.

9. Do you know the reason _____ she got so angry yesterday?

10. I know the boy _____ father is an engineer.

11. _____ is known to all, Taiwan is part of China.

12. I'll show you a store _____ you may buy all _____ you want.

13. That is her mother, _____ is a teacher.

14. That is her mother, _____ she loves very much.

15. The old woman has two sons, one of _____ is a doctor.

II. Find the mistake in each of the following sentences and correct it.

1. The bike I bought it yesterday was broken.

2. I know a girl speaks five languages.

3. The person to who you spoke is very small.

4. The house in that we live is very small.

5. A car which have a big engine is not economical.

6. This is the best film which I have ever seen.

7. Everything which we saw of great interest.

8. His dog, that was now very old, became ill and died.

9. Who want to leave early should sit in the back.

10. He talked of things and persons whom he was interested in.

III Choose the right answer from the four choices.

1. The car ran into a crowd of middle school students, _____ to hospital immediately.
   A. two of whom sent
   B. two of them sent
   C. two of whom are sent
   D. two of them sending

2. He reached London in 1996, _____ , some time later, he became a famous actor.
   A. where
   B. when
   C. which
   D. that

3. —Is that the small town you often refer to?
   —Right, just the one _____ you know I used to work for years.
   A. that
   B. which
   C. where
   D. what

4. I have many friends, _____ some are businessmen.
   A. of them
   B. from which
   C. who of
   D. of whom

5. Mr. Green will come to the party on Sunday, _____ he promised to every one of us.
   A. when
   B. that
   C. where
   D. which

6. She heard a terrible noise, _____ brought her heart into her mouth.
   A. it
   B. which
   C. this
   D. that

7. Finally, the thief handed everything _____ he had stolen to the police.
   A. which
   B. what
   C. whatever
   D. that

8. His parents wouldn't let him marry anyone _____ family was poor.
   A. of whom
   B. whom
   C. of whose
   D. whose

9. The film brought the hours back to me _____ I was taken good care of in that far away village.
   A. until
   B. that
   C. when
   D. where

10. _____ is known to everybody, the moon travels round the earth once every month.
    A. It
    B. As
    C. Which
    D. What

11. This is one of the best books _____ been published so far.
    A. which have
    B. that has
    C. that have
    D. which has

12. She is the only one of the students _____ praised.
    A. who were
    B. that were
    C. which
    D. who was

13. The old man has two daughters, _____ are nurses.

A. two of whom B. both of whom
C. all of them D. neither of them

14. I have five books, _____ is an English book.
    A. all of them B. both of which
    C. one of which D. neither of which

15. Those _____ break the law will be punished.
    A. that B. who
    C. which D. whom

CULTURE TIPS

扫一扫了解如何防范网络诈骗

Unit Six

Virus

Section One Warming Up

Pair work

There may be something wrong with the computer. Match the problems with the Chinese meanings.

1. Hard disk install failure
2. Bad command or file name
3. Drive not ready
4. Insufficient memory
5. Keyboard controller failure
6. Battery is critically low
7. Access denied
8. Cannot find system files
9. Too many open files
10. Maybe infected by virus

A. 电池电量过低
B. 拒绝存取
C. 打开文件太多
D. 可能感染病毒
E. 不能找到系统文件
F. 硬盘安装失败
G. 错误的命令或文件名
H. 键盘控制器出现故障
I. 驱动器未准备好
J. 内存不足

Section Two Real World

Listen and act the following dialogue which is about viruses.

What' wrong with my computer?

Candy: My computer **crashed** when I opened an E-mail **attachment** yesterday.

Rex: It might have been **infected** with a virus.

Candy: Virus? E-mails can carry viruses?

Rex: Viruses most easily **spread** through E-mail attachments. That's why it is essential that you never open E-mail attachments unless you know who it's from.

Candy: So if I open an E-mail with such an attachment, my computer would get infected at once?

Rex: Well, not **exactly**. The E-mail itself is not a virus. A virus must be **attached** to the E-mail to infect an **executable** program. To be more exactly, viruses are executable **files**. Since E-mails are not system files, they cannot infect your computer.

Candy: But the attachment isn't like that, is it?

Rex: Yeah. If you download a file attached to an E-mail and run it, there is a chance that the file may **contain** a virus. You may try some anti-virus software like Kaspersky.

Candy: I see. Thank you very much.

Words & Expressions

crash [kræʃ]	vi. & vt.	fall or strike suddenly, violently, and noisily, especially of things that break 撞碎；崩溃；坠毁
attachment [ə'tætʃmənt]	n.	a supplementary part 附件；附属物
infect [ɪn'fekt]	vt.	put disease into the body(受)传染
spread [spred]	vt. & vi.	become widely known 传播；散布
exactly [ɪg'zæktli]	adv.	Precisely 精确地；确切地
attach [ə'tætʃ]	vt. & vi.	fasten; affix 贴上；附上
executable [ɪg'zekjətəbl]	adj.	capable of being done with means at hand and circumstances as they are 可执行的
file [faɪl]	n.	a set of related records (either written or electronic) kept together 档案；文件
contain [kən'teɪn]	vt.	hold; have within itself 包含；容纳
be attached to		附属于
at once		马上
anti-virus software		杀毒软件

Unit Six Virus

Proper Names

Kaspersky　　　　　　　　　　　　　卡巴斯基

Find Information

Task I Read the dialogue carefully and then answer the following questions.

1. What's Candy's problem?

2. What caused the problem?

3. Should an E-mail attachment be opened if it is from an unknown person?

4. Why can't E-mails infect the computer?

5. What's Rex's advice for Candy at last?

Task II Read the dialogue carefully and then decide whether the following statements are true (T) or false (F).

(　) 1. There's something wrong with Candy's computer when she opened an E-mail attachment.
(　) 2. Viruses are difficult to spread through E-mail attachments.
(　) 3. If people open an E-mail with an attachment carrying viruses, the computer is sure to be infected.
(　) 4. A virus must be attached to E-mail to infect an executable program.
(　) 5. It's a good way to try some anti-virus software to kill viruses.

Words Building

Task I Choose the best answer from the four choices A, B, C and D.

(　) 1. I can't find my keys. I _____ them at the school yesterday.
　　　A. might have left　　　　　　　B. might leave
　　　C. should have left　　　　　　　D. left

(　) 2. It is essential that you _____ every single word correctly.
　　　A. pronounced
　　　B. should pronounce
　　　C. will pronounce
　　　D. pronouncing

( ) 3. _____ you can't answer the question, perhaps we'd better ask someone else.
   A. Since   B. Before
   C. Although   D. Until

( ) 4. —The koala isn't cute, is it? — _____ . I like it very much.
   A. No, it isn't   B. Yes, it is.
   C. Yes, it isn't   D. No, it is

( ) 5. The concert will be held as schedule _____ there is a typhoon.
   A. until   B. if
   C. as soon as   D. unless

Task II Fill in each blank with the proper form of the word given.

1. Please do not send your resume by _____. (attach)
2. Are you sure there is no danger of _____ now? (infect)
3. The volume of this _____ is 2 cubic meters. (contain)
4. What is the _____ size of the room? (exactly)
5. She's _____ kind. (essential)

Task III Match the following English terms with the equivalent Chinese.

A —spyware   1. ( ) 自动升级
B —auto-update   2. ( ) 自动弹启示广告
C —on-demand scanner   3. ( ) 间谍软件
D —system registry   4. ( ) 木马程序
E —rootkit   5. ( ) 恶意广告软件
F —adware   6. ( ) 按需扫描程序
G —pop-up   7. ( ) 混合型威胁
H —Trojan Horse   8. ( ) 后门程序
I —blended threat   9. ( ) 系统注册表
J —backdoor   10. ( ) 黑客软件

Cheer up Your Ears

Task I Listen and write down what you've heard. Then read and recite till you can use them fluently.

1. My computer _____ when I opened an E-mail attachment yesterday.
2. It might have been _____ with a virus.
3. Viruses most easily _____ through E-mail attachments.
4. That's why it is _____ that you never open E-mail attachments unless you know who it's from.

Unit Six   Virus

5. If I open an E-mail with such an attachment, my computer would get infected _____?
6. A virus must be attached to the E-mail to infect an _____ program.
7. To be more _____ , viruses are executable files.
8. Since E-mails are not _____ files, they cannot infect your computer.
9. If you download a file attached to an E-mail and run it, there is a chance that the file may _____ a virus.
10. You may try some _____ software like Kaspersky.

Task II Listen and fill in each blank with the missing word and role-play the conversation with your partner.

A: Hi! Can I help you?
B: It seems all the programs are ___1___. At first, Excel did not work. And then all my files seemed to have gone. Why is my computer acting this way?
A: Let me see. Does the Internet still work?
B: No. I'm ___2___ , but there is.
A: As I can see, you haven't updated your anti-virus protection for three weeks in a row. And the virus is of the ___3___ .
B: Oh, my God! Is there any possibility that my documents get restored?
A: I guess I can handle it. Let's get started.
   (Half an hour later, the problem is solved.)
A: It's OK. You can ___4___ your documents.
B: Many thanks. You do save my computer.
A: If anything is wrong with your computer, don't hesitate to come to our center and we are ___5___ .
B: Thank you.
A: My pleasure.

Task III Listen to 5 recorded questions and choose the best answer.

1. A. You can take a taxi.   B. I'll fly to New York next week.
   C. It's about 20 miles.   D. It's only 600 yuan.
2. A. It's very far from here.   B. It's very small, but nice.
   C. I've got a good idea.   D. I've been there.
3. A. I've got some paper.   B. That's great.
   C. OK, thank you.   D. Oh, no problem.
4. A. Sure. You take the other end. B. Turn right at the next corner.
   C. I think it's too expensive.   D. I'm afraid it can't work.
5. A. I'm sorry to hear that.   B. I like Chinese food.
   C. It's very kind of you to help me. D. I'd like to, but I'll have a meeting.

Table Talk

Task I Complete the following dialogue by translating Chinese into English orally.

A: This computer froze again! Can you come to ___1___ （看一看）my PC? It's acting up again. It must have a virus or something.

B: Just ___2___ （给我点时间）; I'll be right up.

B: I ran a virus scan on your computer, and it turns out that you have a lot of ___3___ （被感染的文件）!

A: But I'm quite careful when I'm browsing the internet. I ___4___ （不知道） how I could have picked up a virus.

B: Well, you have to ___5___ （确保） that your anti-virus software is updated regularly; yours wasn't up to date, that's probably what was causing your problems.

A: Ok. Anything else?

B: Yeah, try not to kick or hit the computer!

A: Um yeah. Sorry about that.

Task II Pair-work. Role-play a conversation talking about viruses with your partner.

Situation

Imagine your computer doesn't work because of the viruses. You want to ask your friend for some advice. Now you are talking with your friend.

Section Three Brighten Your Eyes

Pre-reading questions:

1. Has your computer been infected by viruses? How do you deal with that situation?
2. How can viruses affect your computer?

Impact of Viruses on Computer Systems

Computer viruses, as we know now, **originated** in 1986 with the **creation** of Brain-the first virus for personal computers. Two brothers wrote it (Basit and Farooq Alvi who ran a small software house in Lahore, Pakistan) and started the race between viruses and anti-virus programs which still goes on today. Virus can be **reprogrammed** to do many kinds of **harm** including the following.

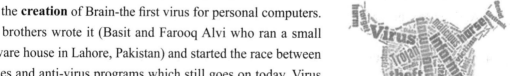

- → Copy themselves to other programs or areas of a disk.
- → **Replicate** as **rapidly** and **frequently** as possible, filling up the infected system's disk and memory **rendering** the systems useless.
- → **Modify**, **corrupt** or destroy **selected** files.
- → **Erase** the contents of entire disks.
- → Lie **dormant** for a **specified** time and until a given condition is met, and then become active.
- → Open a back door to the infected system which allows someone else to access and even control the system through a network or internet connection.
- → Steal important information of the system or the user and send it to a specified location.

Words & Expressions

impact [ˈɪmpækt]	n./v.	strong effect on sb/sth 影响
originate [əˈrɪdʒɪneɪt]	vt. & vi.	have a cause or beginning 发源于；来自
creation [kriˈeɪʃn]	n.	the act of starting something for the first time 创造
reprogramme [rɪpˈrəʊɡræm]	v.	rewrite a computer program 再次（重新）设定程序
harm [hɑːm]	n.	damage or wrong 危害；伤害
	vt.	cause harm to (使)受到损害；伤害
replicate [ˈreplɪkeɪt]	vt.	reproduce or make an exact copy of 复制
rapidly [ˈræpɪdli]	adv.	with rapid movement 迅速地
frequently [ˈfriːkwəntli]	adv.	many times at short intervals 频繁地
render [ˈrendə(r)]	vt.	cause to be in some condition 使；致使
modify [ˈmɒdɪfaɪ]	vt. & vi.	change 修改；更改
corrupt [kəˈrʌpt]	v.	to cause mistakes to appear in a computer file, etc. 引起（计算机文件等的）错误；破坏
selected [sɪˈlektɪd]	adj.	chosen in preference to another 选定的
erase [ɪˈreɪz]	vt.	rub out or remove sth. 擦去；抹掉
dormant [ˈdɔːmənt]	adj.	inactive but capable of becoming active 休眠的；潜伏的
specified [ˈspesɪfaɪd]	adj.	clearly and explicitly stated 指定的；规定的
as … as possible		尽可能……
fill up		装满；堵塞
given condition		特定条件

Find Information

Task I Read the passage carefully and then answer the following questions.

1. When did computer viruses appear?
   _____

2. What's the name of the first virus for personal computers?
   _____

3. How many kinds of harm are mentioned in the passage?
   _____

4. Is a computer virus active from the beginning?
   _____

5. How can some viruses crash the system?
   _____

Task II Read the passage carefully and then decide whether the following statements are true (T) or false (F).

(　) 1. Two brothers created the first virus for personal computers.
(　) 2. Viruses can copy themselves to other programs or areas of disk.
(　) 3. Viruses can only erase some contents of the disks.
(　) 4. Viruses are active all the time.
(　) 5. The information of the system can be stolen by viruses.

Words Building

Task I Translate the following phrases into English.

1. 个人电脑_____
2. 杀毒程序_____
3. 磁盘和内存_____
4. 选定的文件_____
5. 清除内容_____
6. 潜伏_____
7. 访问系统_____
8. 窃取信息_____

Task II Translate the following sentences into Chinese or English.

1. Computer viruses, as we know, do many kinds of harm to the system.
   _____

2. Computer viruses lie dormant for a specified time until a given condition is met, and then become active.
   _____

3. All computer viruses can replicate themselves by making a copy of themselves over and over again.
   _____

4. You should scan your computer system as frequently as possible.
   _____

5. To protect your computer system, you can use an anti-virus program.
   _____

6. 第一个个人电脑病毒起源于 1986 年。
   _____

7. 任何电脑都能很容易地被病毒感染。
   _____

8. 计算机病毒可以窃取系统重要信息并发送到指定位置。
   _____

9. 现在病毒比以前传播地更快。
   _____

10. 病毒侵占受感染的系统磁盘和内存并致使系统无法工作。
    _____

Task III **Choose the best answer from the four choices A, B, C and D.**

(　) 1. Tell Mary that there's someone _____ for her at the door.
   A. waiting   B. waited
   C. waits   D. to wait

(　) 2. There is a big dog _____ to a fence outside the house.
   A. tying   B. tied
   C. to tie   D. ties

(　) 3. Is this the factory _____ some foreign friends visited last Friday?
   A. that   B. where
   C. who   D. the one

(　) 4. My parents don't allow me _____ out at night.
   A. go   B. going
   C. to go   D. goes

(　) 5. Some of the plastic bags can't _____ after June 1.
   A. use   B. be use   C. be used   D. are used

Task IV **Fill in each blank with the proper form of the word given.**

1. Freezing weather is _____ to orange trees. (harm)

2. He's a _____ visitor to our house.(frequently)

3. His success is _____ due to hard work.(entire)

4. The police tried to _____ him with the murder. (connection)

5. Confucius is the _____ of Confucianism. (originate)

Cheer up Your Ears

Task I Listen and write down what you've heard. Then read and recite till you can use them fluently.

1. Virus can do many kinds of _____ to computer systems.
2. If the users are _____ traditional viruses, they should run a more secure operating system like UNIX.
3. Automatic protection of anti-virus software should be _____ at all times.
4. Viruses can replicate as rapidly and _____ as possible.
5. Until a given condition is met, viruses will become _____.
6. Users should buy legal _____ of all software.
7. The users should _____ a manual scan of their hard disks weekly.
8. Newer viruses can send email _____ that appears to be from a personal user.
9. The first virus for personal computers _____ in 1986.
10. Appropriate _____ should be assigned to the shared network drives.

Task II Listen to 5 short dialogues and choose the best answer.

1. A. He has refused.　　　　　　　　B. He is glad to help.
   C. He doesn't have license.　　　　D. He can't drive a car.
2. A. The children are enjoying themselves.
   B. The children didn't come to the party.
   C. The children are giving a lot of fun in the party.
   D. The children are not behaving themselves in the party.
3. A. The woman's car is not worth repairing.
   B. The woman can get her car in a short while.
   C. He's not ready to repair the woman's car.
   D. Repairing a car is very expensive.
4. A. To type the paper for the woman.
   B. To visit the woman.
   C. To copy the woman's essay.
   D. To read the woman's paper.
5. A. He likes his job very much.
   B. He dislikes his job, but has to do it.
   C. He doesn't take the job seriously.
   D. He feels the work too heavy.

Table Talk

Group Work

Talk about the impact of viruses on computer systems with your partner as much as you know.

_____

_____

Section Four Translation Skills

科技英语的翻译方法与技巧——词汇的增译

由于英语和汉语在遣词造句方面的差别,翻译时应注意在词量上有所增减。有时需要在译文中增加原文中无其形而有其意的词。有时原文中的有些词在译文中译不出来,因为译文中虽无其词,但已有其意。增译法主要用于英语省略结构、短语结构、有指代关系的结构和有动作意义的名词。

1. 增译语气连惯性的词

为使译文的措辞连贯,要增译语气连贯性的词。例如:

This question is really a technique design rather than an operation problem.

这个问题实际上是属于工艺设计方面而非操作方面的问题。

2. 英语中表达动作意义名词的增译

在翻译英语文献时,根据上下文的意义补充一些表达动作意义的名词,如"作用""现象""效应""方法""过程"等。例如:

Correction and changes are made in accordance with the observed performance.

根据观察到的结果进行修正和替换。

3. 英语复数名词的增译

复数名词前可增加恰当的表示复数概念的词,如"一些""许多""一批""各种"等词,使其复数意义更加明确。例如:

After a series of experiment important phenomena have been ascertained.

经一系列实验后,查明了一些重要现象。

4. 增译解说性的词

增加反应背景情况的词、解说性的词以使汉语语义明确,根据上下文增词,符合业内人士的表达习惯。例如:

Know where the emergency stop button is before operating the lathe.

操作机床前要知道紧急按钮的位置。

这里 emergency stop button 增译"按钮"二字。

Keep tools overhang as short as possible.

在不影响操作的情况下，刀具伸出部分越短越好。增译"在不影响操作的情况下"使语义更加严谨。

The constant temperature workshop consists of large parts machine line, middle and small shell parts machine line, shaft and plate parts machine line, precision machine line, unit assembly line, powder spraying line, final assembly line, automatic solid storehouse, and precision quality test.

恒温车间建有大件加工线、中小壳体类零件加工线、轴类及盘类零件加工线、精密加工线、部件装配线、涂装作业线、总装作业线、全自动立体仓库、精密检测室。根据句子意思，precision quality test 应增译为精密检测室。

The more complex the casting, the more difficult the alloy; the more difficult the application specifications, the greater the financial benefit.

铸件结构越复杂，合金越难熔炼；铸件使用的技术要求越高，那么经济效益就越大。

With MAZAK CAMWARE, MAZATROL programs can be generated by a PC. With the Cyber Tool Management of CPC, tool managers can monitor the state of the tools for on line machines and prepare spare tools for replacing worn tools.

通过 CAMWARE 模块，MAZATROL 加工程序由计算机生成。通过 CPC 的智能刀具管理模块，刀具管理人员可以监控在线机床的刀具使用状态，准备备用刀具，及时更换超出寿命期的磨损刀具。根据上下文，增加"模块"二字。

5. 抽象名词后加名词使其具体化

翻译某些动词或形容词派生来的抽象名词时，可根据上下文增添适当的名词，使其更符合汉语习惯。如增译"功能""效应""现象""方法""装置""变化"等。

A feedback device at the opposite end of the ball screw allows the control to confirm that the commanded number of rotations has taken place.

安装在滚珠丝杠另一端的反馈装置使数控装置能确认，实际上是否转动了被指令的转数。

On machines with the TSC option, clean the chip basket on the coolant tank.

对于带主轴中心孔冷却（TSC）选择功能的机床，清理冷却箱上的切屑收集篮。

6. 增加表达时态的词

英语动词靠词形变化（go, went）或加助动词（will go, have gone）来表达。汉语动词没有词形变化，英译汉时要增加汉语特有的时态助词或表示时间的词，如"曾、了、过、着、将、会"等。此外，为了强调时间概念或时间上的对比，往往也需要增加一些其他的词。

Carbide tipped tools will stand speeds in excess of those recommended for high-speed steel tools.

硬质合金刀具可以承受表中推荐的基于高速钢刀具的转速。

7. 增译概括性的词

Both steel iron and cast iron have the structures of eutectoid, hypoeutectoid and hypereutectoid.

钢和铸铁都有共析、亚共析和过共析 3 种组织。

8. 连词的减译

连词在英语中用的比较广泛，汉语中连词用的较少。大多数汉语句子是按时间顺序和逻辑顺序排列的，各成分之间的关系清楚。在英译汉时，可以将英语中的某些连词减译。例如：

These metals require a certain amount of preheating before welding and then allowed to cool slowly after the weld is completed.

这些金属通常要求在焊接前进行一定的预热，在焊接完成后缓慢冷却。

9. 增添原句中省略的成分

英语倒装句中为了避免句子部分内容重复而省略的部分，译成汉语时应把它表达出来。

Listening

Task I **Listen to 2 conversations and choose the best answer.**

Conversation 1

1. A. At the Customs.　　　　　　　　　　B. At an information desk.
   C. At a luggage claim area.　　　　　　D. At the waiting hall.
2. A. Oxford.　　　　　　　　　　　　　　B. Cambridge.
   C. London.　　　　　　　　　　　　　　D. Birmingham.
3. A. One.　　　　　　　　　　　　　　　B. Two.
   C. Three.　　　　　　　　　　　　　　D. Four.

Conversation 2

4. A. Having a meal in a restaurant.
   B. Checking in at a hotel.
   C. Booking a ticket at a station.
   D. Reserving a seat in a cinema.
5. A. $ 3.00　　　　　　　　　　　　　　B. $ 13.00
   C. $ 30.00　　　　　　　　　　　　　D. $ 33.00

Task II **Listen to the passage and fill in the blanks with the missing words or phrases.**

Almost every activity in life requires communication. When you make a speech at school, ___1___ your food at a restaurant, or tell a joke, you are communicating. Learning to speak and express your thoughts ___2___ is the basic requirement of communication. But there is ___3___ to speak which we often neglected. That's listening. If you don't listen, how do you know what to say when your workmate needs help or your friend is upset? So, try to remember that good communications listen twice ___4___ as they speak. Maybe that's why God gave us two ears and just one ___5___.

Extensive Reading

Directions: After reading the passage, you are required to complete the outline below briefly.

Viruses often perform some type of harmful activities on infected hosts, here are some tips for protecting your personal computer.

→ Always keep your operating system and programs up-to-date.

→ Install an anti-virus and anti-spyware program that automatically scans for viruses when the system boots.

→ Update the virus definitions daily to ensure your system is protected against the latest threats.

→ Do not download any files from the Internet unless you are certain the source is not transmitting a virus to you.

→ Do not use any storage media that has been used in another computer unless you are certain the other computer is free of viruses and will not pass the virus on to your system.

→ Never open e-mail attachments from people you don't know; and don't open any file attachment that ends in ".exe". All downloaded files should be scanned by your anti-virus application before you run or install it.

How to Protect Your Computer against Virus

① Always keep your operating system and programs ___1___.
② Install an anti-virus and ___2___ that automatically scans.
③ Update the virus definitions ___3___.
④ Do not ___4___ any files from the Internet if you're not sure it's safe.
⑤ Do not use any ___5___ that has been used in another computer unless you are certain the other computer is ___6___ of viruses
⑥ Never open e-mail ___7___ from people you don't know.
⑦ All downloaded files should be ___8___ by your anti-virus application before you run or install it.

Section Five Writing

Reports 报告

报告一般是下级向上级呈报的，针对某一事由或某一主题，在开展调查研究的基础上得出结论后所汇报的内容。根据内容不同，可分为调查报告、工作报告、情况报告、建议报告、答复报告和递送报告等。正式报告较长，内容较多；非正式报告则短，但在实际工作中的使用却更频繁。

Sample

Proposal Reports
To: General Manager
From: Public Relations Manager
Date: 30th August, 2007
Subject: Report on the English Standard of Our Promotional Materials
Introduction
According to your instructions of 26th July 2007, to report on the English standard of our

promotional materials, including brochures, leaflets, speeches and advertisements and to make recommendation.

Findings

150 pieces of promotional materials were examined. They include 50 brochures, 40 leaflets, 30 speeches and 30 advertisements. Each piece of materials was assigned a grade according to their content, language readability and grammatical accuracy.

Only 20 pieces of articles were classified as excellent, 40 pieces as good and 60 as acceptable, the rest 30 were regarded as below standard and unclassified.

Conclusion

It is clear that the English proficiency of our marketing staff is far from satisfactory.

Based on the above findings, I recommended that we should:

Set up a special course designed to develop writing skills for all marketing staff. Special emphasis should be placed on promotional materials.

建议报告

致：总经理

由：公关经理

日期：2007年8月30日

主题：关于提高宣传品英语水平的研究报告

引言

根据您2007年7月26日的指示，我完成了对公司宣传品，包括小册子、传单、演讲词及广告的语言评估工作，并做出相应建议。

结果

我一共审阅了150份宣传品，包括50份小册子、40份传单、30个演讲词及30个广告。每份文件均根据内容、可读性及文法给予一个等级。

只有20篇文章被列入优等，40篇属良好，60篇可以接受，其余30篇为不够标准的文章。

结论

很明显，营销人员的英文程度并不理想。

根据以上结果，做如下建议：

为营销人员开设一门提高写作技能的专门课程，特别着重宣传品的撰写方法。

写作格式：

报告一般分为4个部分：扉页、引言、正文（结果）和结论。

扉页：包括报告的题目、收阅人、撰写人、日期等。

引言：通常会介绍报告的目的、背景、所报告的问题和所采用的方法等。

正文：介绍调查研究的具体实施情况、研究的发现和结果的分析、得出的结论等。

结论：这是报告的结尾部分，是对于所发现的结果的总结，并提出相应的意见和建议。

报告格式如下表所示：

To（呈送）：
From（发自）：
Date（日期）：
Subject（主题）：
Instruction（引言）
Findings（正文）
Conclusion（结论）

语言特点：
目的明确，语言简单明了，客观而富有逻辑性，内容详尽，重点突出，条理清晰。

常用句型：
1. Upon the request of …, we have inspected…
   应……要求，我们调查了……
2. In order to improve…, it is hereby recommended that…
   为了提高……，因此建议……
3. Here is the report about the work done by the marketing department of March 2007.
   这是我们营销部对2007年3月的工作情况所做的报告。
4. The purpose of this report is to improve efficiency.
   本报告的目的是为了提高效率。
5. Work in progress is as follows:
   目前正在进行的工作如下：

Practice

You are required to fill in a Report according to the following information in Chinese. You should include all points in the following table.

To: Mr. Wang Jin, President of Shanghai University
From: Zhang Li, Dean of studies
Date: May 6, 2006
Subject: _____（购买计算机和摄像机）
Upon the request of the Equipment Division of the University, we have inspected the lab of the Mathematics Department and found its present equipments unsatisfactory to students, particularly to postgraduates. _____（为了改进实验效果，兹特建议购买30台计算机和2台摄像机） and issued to the laboratory.

Section Six Grammar

Subjunctive Mood I 虚拟语气（一）

英语中动词除了有时态（tense）和语态（voice）的变化之外，还有语气（mood）的变化。语气表示说话者的意图、看法或态度。英语中有3种语气，即陈述语气、祈使语气和虚拟语气。

陈述语气（indicative mood）表示所说的话是事实。

祈使语气（imperative mood）表示所说的话是请求或命令等。

虚拟语气（subjunctive mood）表示所说的话不是事实，而是与事实相反的假设或是不可能发生的情况，或是一种主观愿望、建议、要求等。

有时为了客气、委婉，也使用虚拟语气。

要用好虚拟语气，除了先弄清楚何时该用虚拟语气外，还必须牢记：如何虚拟，即虚拟表达方式。通常来说，虚拟语气多用于从句中，表现为动词的形式变化。大体上可归纳为两种虚拟方式：（时态变化型 和 should 型）。首先，让我们学习时态变化型虚拟。

请牢记： 时态要在原有基础上向过去倒退一格。

通过时态倒退法来表达虚拟语气的情况有：

1. wish（希望、但愿） 后接宾语从句表示不可能实现的愿望。例如：

（1）I wish (that) I had been to every country in the world.
我希望我已去过世界上的每一个国家。

（2）I wish (that) I were twenty years younger. 我希望我年轻20岁。

☆（注：在虚拟语气中，常用 were 代替 was。）

2. as if / as though（好像） 后接状语从句表示不符合实际的主观想象。例如：

（1）She treats me very well as if she were my own mother.
她对待我很好，就好像她是我的亲生母亲一样。（实际上，她不是。）

（2）He often behaves as though he were better than us.
他经常做出一副比我们都强的样子。（实际上，其实他并不比我们强。）

（3）Tom talked about Paris as though he had been there himself.
汤姆谈起巴黎就仿佛他到过那里似的。（实际上，汤姆并没到过巴黎。）

3. if only（要是……就好了） 后接从句 = How I wish…例如：

If only I were a bird. 要是我是一只鸟就好了。（事实上，我不可能成为鸟。）

If only I could fly. 要是我会飞就好了。（实际上，我不会飞。）

If only I hadn't lost my wallet. 要是我钱包没丢就好了。（实际上，钱包已丢。）

4. would rather（宁愿） 后接宾语从句要虚拟，但时态变化需注意：

☆ 表示与现在和将来事实相反的虚拟语气要用动词的过去式（did）

☆ 表示与过去事实相反的虚拟语气用过去完成式（had done）例如：

I would rather she were not here now. 我宁愿她现在不在这里。

I would rather you came tomorrow. 我宁愿你明天来。

I didn't sleep well. I'd rather we hadn't seen the scary film last night.
我没睡好。我宁愿我们昨夜没有看那个可怕的电影。

（实际上，我看了恐怖电影后没睡好。）

5. if（如果）引导的虚拟句表示不可能实现的假设。 注意：

☆主句时态和从句时态都要在原有基础上向过去倒退一格。

☆if 虚拟要分清时间，区别现在、过去和将来。

具体变化形式见下表：

虚拟语气	从句	主句
与现在事实相反	一般过去时（did）（系动词用 were）	would/should/could/might+do
与过去事实相反	过去完成时（had done）	would/should/would/might+have done
与将来事实相反	（1）一般过去时（did） （2）were to do （3）should do	would/should/would/might+do

例：（1）If he were here now, he might be able to help you.

虚拟现在：如果他此时在这里，他就能帮你了。

（事实是：他现在不在这，不可能帮你。）

（2）If you had got up earlier yesterday, you would have caught the train.

虚拟过去：如果你昨天早起些，就会赶上火车的。

（事实是：你昨天没早起，没赶上车。）

（3）She would have gone to the party if she had been invited.

虚拟过去：要是当初她受邀的话，她就会去参加那次聚会了。

（事实是：当时没请她，她没去。）

（4）If she hadn't called me, I would have overslept this morning.

虚拟过去：今早，她要是不叫我的话，我就会睡过头了。

（事实是：她叫我了，我没睡过头。）

（5）（It's summer now）If it snowed tomorrow, I would make a snowman. 虚拟将来：如果明天下雪的话，我将做个雪人。

（事实是：夏天不可能下雪，从句可用两种方式，主句 would do…）=If it should snow tomorrow, I would make a snowman.= If it were to snow tomorrow, I would make a snowman.

☆重点：学会区分由 if 引导的真实条件句和虚拟句（二者时态不同）

if 虚拟句的时态都跟过去有关；真实条件句通常为"主将从现"。例如：

It's winter now. If it snows tomorrow, I will make a snowman.

（此句为真实条件句，因为冬天下雪有可能发生。故时态正常："主将从现"。）

此外，if 虚拟句还应了解和掌握以下情况：

☆ A. 省略 if 的虚拟句

当 if 引导的非真实条件句中的谓语部分有 **were, had, should** 时，可以把 if 省略，将 were, had, should 提到主语的前面，构成倒装结构。例如：

（1）Were you in my position, you would do the same.

=If you were in my position,… 假如你处在我的地位，你也会这样干的。

（2）Had you not helped me, I whould have failed.

=If you hadn't helped me,… 要是没有你的帮助，我就失败了。

（3）Should they attack us, we would wipe them out completely.

=If they should attack us,… 假如他们进攻我们，我们就把他们彻底消灭。

= Were they to attack us, …

= If they were to attack us, …

☆ B. 混合型的条件句 =（错综时间条件句）

当 if 条件从句与主句的时间不一致时，主句和从句的谓语动词形式不必按照上表相互呼应，而要将虚拟语气的形式按各自的时间做相应的调整。这种条件句叫混合条件句或错综时间条件句。较多见的是从句虚拟过去（用 had done），而主句虚拟现在（用 would/should/might/ could + do）。例如：

（1）He would be better now if he had taken my advice.

如果他听了我的劝告，现在他就会更好些。

（2）If I had spoken to him yesterday, I should know what to do now.

假如昨天我对他说了，现在我就知道该怎么办了。

☆ C. 含蓄条件句

虚拟条件句中的条件有时不直接表达出来，而是暗含在关键词、短语及上下文中，叫含蓄条件句。常见标志为 without, but for（要是没有），or , otherwise（否则）等。例如：

（1）But for your help, we couldn't have succeeded in the experiment.

如果没有你的帮助，我们的实验是不会成功的。

But for …= If it had not been for… (要是没有…) 虚拟过去

（2）I didn't have enough money. **Otherwise**, I would have bought the computer yesterday.

昨天我的钱不够。否则的话，我会买下那台电脑的。

Exercises

I. Fill in the blanks with the proper forms of the words given in brackets.

1. If he _____out, I'll call tomorrow. (be)

2. If I _____in your position, I would make a different choice. (be)

3. If I had seen him yesterday, I _____it to him. (give)

4. If you had taken my advice, you _____much better now. (feel)

5. But for your timely help, the little boy _____his life. (lose)

6. —Have you been to New York?—No, but I wish I _____(have).

7. He is talking so much about Thailand as if he _____there many times. (be)

8. If you _____me, you wouldn't get upset at every little thing I say! (love)

9. _____I you, I would quit smoking. (be)

10. Sally is crying sadly. I'd rather you _____her the bad news just now.(not tell)

II. Choose the best answer

(　) 1. If only I _____ as young as you are!

　　　A. being　　　　　　　　　　　　　　B. am

C. be   D. were

( ) 2. If you _____ five minutes earlier, you _____ him.
　　　A. should come, had seen　　　　　B. came, would see
　　　C. come, will see　　　　　　　　　D. had come, would have seen

( ) 3. When a pencil is partly in a glass of water, it looks as if it _____ .
　　　A. breaks　　　　　　　　　　　　B. was broken
　　　C. were broken　　　　　　　　　　D. had been broken

( ) 4. _____ your letter, I would have written back two days ago.
　　　A. If I received　　　　　　　　　　B. If I receive
　　　C. Had I received　　　　　　　　　D. If I have received

( ) 5. I wish I _____ you yesterday .
　　　A. seen　　　　　　　　　　　　　B. did see
　　　C. had seen　　　　　　　　　　　 D. were to see

( ) 6. _____ this afternoon, he would get there by Friday.
　　　A. Would he leave　　　　　　　　 B. Was he leaving
　　　C. Were he to leave　　　　　　　　D. If he leaves

( ) 7. We could not have succeeded _____ your help.
　　　A. but for　　　　　　　　　　　　B. without
　　　C. if it had not been for　　　　　　D. all the above

( ) 8. _____ I be free tomorrow, I could go with you.
　　　A. Could　　　　　　　　　　　　 B. Should
　　　C. Might　　　　　　　　　　　　 D. Must

( ) 9. He looked as if he _____ ill for a long time.
　　　A. was　　　　　　　　　　　　　B. Were
　　　C. has been　　　　　　　　　　　D. had been

( ) 10. I left very early last night, but I wish I _____ so early.
　　　A. didn't leave　　　　　　　　　　B. hadn't left
　　　C. haven't left　　　　　　　　　　D. couldn't leave

III. Try to find the mistakes and correct them.

1. They talked about the city as if they have been there.
2. If he came earlier, he would have met my brother.
3. He will come tomorrow. —But I'd rather he will come next week.
4. How I wish I haven't lost my money.
5. Were it rain tomorrow, we would put off the football game.

IV. Rewrite the sentences.

1. I can't fly, but I wish to fly. =If only I _____ _____ .
2. If it hadn't been for your help, I might have failed.
　 = _____ _____ your help, I might have failed.
3. As he is in poor health, he won't go abroad with us.
　 If he _____ in good health, he _____ _____ abroad with us.

4. He talks like a baby. = He talks as if he _____ a baby.
5. Sarah looks at her husband. It seems that he is a stranger.
   Sarah looks at her husband _____ _____ he were a stranger.
6. It is cold outside. We don't let the children play in the garden.
   If it _____ not cold outside, we would let the children play in the garden.
7. If I were at school again, I would study harder.
   _____ _____ at school again, I would study harder.
8. Should it rain tomorrow, we would not go climbing.
   _____ _____ _____ rain tomorrow, we would not go climbing.
9. If I had known it, I'd have told you.
   _____ _____ _____ it, I would have told you.
10. If we hadn't had computers, we would not have achieved much of today's advanced technology.
    _____ computers, we would not have achieved much of today's advanced technology.

V. Fill in the blanks and finish the dialogue.

A: I miss my folks. I wish we __1__ (can) see them.
B: Well, if they __2__ (live) closer, we would see them more often.
A: If we owned a bigger house, they __3__ (can) live with us.
B: You know they would never do that. They'd feel that they were intruding.
A: You're right. But I wish I __4__ (be) with them now. Let's go to visit them.
B: We could if we didn't have to work and if the kids didn't have school.
A: Maybe they __5__ (come) to visit us when it gets warmer.
B: I'm sure they will. Maybe we can go to see them during this summer vacation.
   Why don't you call them on the phone?
A: I did just now. That's why I miss them so much right now.

CULTURE TIPS

扫一扫认识常见病毒

Unit Seven

Website

Section One Warming Up

Pair work

Talk about the following websites.

Section Two Real World

Listen and act the following dialogue which is about creating a website between a customer and a tech support.

I want to create a website for my company

Candy: Good morning, sir. Can I help you?

Rex: Yes. I'm thinking about creating a website for my company.

Candy: That's cool.

Rex: What should I do?

Candy: First, you should set up a **domain**. Domain **registration** is easy and inexpensive.

Rex: Which domain name is the best?

Candy: The dotcom domain is **recommended** unless you are **targeting** a **specific geographic** market such as the UK or Germany or another country.

Rex: Oh, the dotcom domain is OK. What's next?

Candy: Choose paid hosting services, which is around the cost of a cup of Starbucks Coffee.

Rex: Are there any free hosting services?

Candy: Yes. But you may be **forced** to display **advertising**. Also, free services are usually less **reliable** and slower. And then, you can build your site.

Rex: You mean the content and design?

Candy: Yeah. You don't need to have a high **tech** website with all the latest **widgets** in order to have a useful site. Of course if you plan to run a **forum**, you will need more than just **text**.

Rex: OK. Is that enough?

Candy: Having a website means very little if nobody goes there. You need to offer valuable and **attractive** information to get a lot of visitors to your website.

Rex: It seems there is much to create a website. Thanks for your help.

Candy: My pleasure.

Words & Expressions

domain [dəˈmeɪn]	n.	particular field of thought, activity, or interest, especially one over which someone has control, influence, or rights 范围；领域
registration [ˌredʒɪˈstreɪʃn]	n.	the act of enrolling 注册
recommend [ˌrekəˈmend]	v.	speak to sb. in favour of; praise as being good for a purpose 推荐
target [ˈtɑːgɪt]	vt.	intend (something) to move towards a certain goal 把……作为目标

specific [spəˈsɪfɪk]	adj.	particular; fixed 具体的；特定的
geographic [ˌdʒiːəˈgræfɪk]	adj.	of or relating to the science of geography 地理的
force [fɔːs]	vt.	make (an unwilling person) doing sth 强制；迫使
advertising [ˈædvətaɪzɪŋ]	n.	a public promotion of some product or service 广告
reliable [rɪˈlaɪəbl]	adj.	dependable（形容词）可靠的
tech [tek]	n.	the practical application of science to commerce or industry 技术
widget [ˈwɪdʒɪt]	n.	a device or control that is very useful for a particular job 小部件
forum [ˈfɔːrəm]	n.	any place where public matters may be talked and argued about; a meeting for such a purpose 论坛；讨论会
text [tekst]	n.	the main body of writing 文本；正文
attractive [əˈtræktɪv]	adj.	having power to arouse interest 有吸引力的
set up		建立
such as		例如
hosting service		主机托管业务
in order to		为了，目的是

Proper Names

Starbucks Coffee   星巴克咖啡

Find Information

Task I Read the dialogue carefully and then answer the following questions.

1. What does Rex want to do?
   _____

2. Which domain name is recommended?
   _____

3. Is the cost of hosting services very high?
   _____

4. What's the disadvantage of free hosting services?
   _____

5. What should be done to get a lot of visitors?
   _____

Task II Read the dialogue carefully and then decide whether the following statements are true (T) or false (F).

(　) 1. It's easy and inexpensive to set up a domain.
(　) 2. There's only one domain name.
(　) 3. All hosting services are paid.
(　) 4. Free hosting services may force your website to display advertising.
(　) 5. Only text is needed to run a forum on the website.

Words Building

Task I Choose the best answer from the four choices A, B, C and D.

(　) 1. A fund will be _____ for the dead men's families.
　　A. given up　　　　　　　　　　B. set up
　　C. cut up　　　　　　　　　　　D. put up
(　) 2. Circumstances force us _____ this policy.
　　A. adopt　　　　　　　　　　　B. to adopt
　　C. adopting　　　　　　　　　 D. adopted
(　) 3. I don't like this one. Please show me _____ .
　　A. other　　　　　　　　　　　B. others
　　C. another　　　　　　　　　　D. the other
(　) 4. The experiment was _____ easier than we had expected.
　　A. more　　　　　　　　　　　B. much
　　C. a lot of　　　　　　　　　　D. many
(　) 5. Can you _____ me some new books on this subject?
　　A. demand　　　　　　　　　　B. recommend
　　C. command　　　　　　　　　 D. mend

Task II Fill in each blank with the proper form of the word given.

1. I bought this magazine on the _____ of a friend. (recommend)
2. You must _____ if you intend to vote. (registration)
3. Of all the subjects, I like _____ best. (geographic)
4. The writing of poems, stories or plays is called _____ writing. (create)
5. Try to be a man of great _____ . (valuable)

Task III Match the following English terms with the equivalent Chinese.

A—sign in　　　　　　　　1. (　) 网址
B—web designer　　　　　 2. (　) 主页

Unit Seven Website

C —terms of use
D —web address
E —hyperlink
F —business card
G —home page
H — back to top
I — web page layout
J —confirm password

3. (　) 页面设计
4. (　) 再次确认密码
5. (　) 登录
6. (　) 名片
7. (　) 网页设计员
8. (　) 使用条款
9. (　) 返回顶部
10. (　) 超链接

Cheer up Your Ears

Task I Listen and write down what you've heard. Then read and recite till you can use them fluently.

1. First you should set up a _____.
2. Domain registration is easy and _____.
3. The dotcom domain is _____ unless you are targeting a specific geographic market.
4. Choose _____ hosting services.
5. You may be _____ to display advertising.
6. Free services are usually less _____ and slower.
7. You don't need to have a _____ tech website with all the latest widgets.
8. If you plan to run a forum, you will need _____ just text.
9. Having a website means very _____ if nobody goes there.
10. You need to offer _____ and attractive information to get a lot of visitors to your website.

Task II Listen and fill in the blanks with the missing words and role-play the conversation with your partner.

A: Hi, I'm making a __1__.
B: Really?
A: Yeah. Having a web address is almost __2__ having a street address.
B: What's on your website?
A: At the top of the pages is a graphic banner. Next comes a greeting and a short __3__ of the site. Pictures, texts, and links to other websites follow.
A: That sounds cool.
B: Don't you want one?
A: Yeah, but it's __4__?
B: Yes and no. Creating pages takes a little practice, but once you get the hang of it, it's __5__.

Task III Listen to 5 recorded questions and choose the best answer.

1. A. It's open at 9 a.m.　　　B. Sorry, I have no idea.
   C. That's all right.　　　　D. Thank you.
2. A. She's an English student.　B. She's interested in music.
   C. She's a friend of mine.　　D. She's tall with dark hair.
3. A. Thanks.　　　　　　　　B. I don't think so.
   C. Oh, no.　　　　　　　　D. It doesn't matter.
4. A. Sandwich and coffee.　　B. Beer, please.
   C. It's my favorite food.　　D. I don't like this meal.
5. A. A good one.　　　　　　B. One dollar.
   C. Six times a year.　　　　D. In a book store.

Table Talk

Task I Complete the following dialogue by translating Chinese into English orally.

A: Hello! It's you, Jane? Are you free these days?
B: Sorry, I've been busy lately.
A: Doing what?
B: I've been ___1___ （创建一个网站）.
A: Have you got it done?
B: Not yet. I think it will be completed ___2___ （一周后）.
A: You are really somebody. But what's the use of it?
B: I want to ___3___ （与别人交流）just about anything.
A: But how can you make it?
B: Through the link sharing, anyone can find my page and ___4___ （与我联系）.
A: Oh, I see. So if I could, I would like to have one.
B: Of course you can. It's not difficult.
A: ___5___ （感谢）your encouragement.

Task II Pair-work. Role-play a conversation about creating a website with your partner.

Situation

Imagine you are going to create a website. You want to ask your friend for some advice. Now you are talking with your friend.

Section Three Brighten Your Eyes

Pre-reading questions:
1. What's your favorite website? What do you learn from the website?
2. What kind of website can attract you most?

Website Design Tips

When designing your website, you should follow some basic rules. These tips make web surfing easier for your visitors.

Tip 1 Design your website for your visitors.

You should know what they are looking for before you ever write one word for your page. Make your site attractive to them by adding information that they are looking for.

Tip 2 Make your website easy to read.

The website can be made worse by bad **font** and color choices for your text.

Tip 3 Your website should be easy to **navigate**.

You can use either **graphic** images, such as buttons and **tabs**, or text **links**. Whichever you use, make sure they are **labeled** clearly and **accurately** and **familiar** to your visitor. Make sure that all your pages have a link back to your main page.

Tip 4 Make your website easy to find.

The most common way to **promote** a website is through search engines, **directories**, **award** sites, emails as well as links from other websites. Put your **contact** information on your website in case someone needs to contact you.

Tip 5 Keep your web page **layout** and design **consistent**.

Tip 6 Your website should be quick to display.

Avoid using too many graphics and **animations** as these files eat up **bandwidth** with your server. Your page should be displayed fully in the first 10 seconds or else your visitor will feel tired and **frustrated**.

Words & Expressions

font [fɒnt]	n.	a specific size and style of type within a type family 字体
navigate ['nævɪgeɪt]	vi. & vt.	direct carefully and safely 航行；操纵
graphic ['græfɪk]	n.	an image that is generated by a computer 图表
tab [tæb]	n.	the key on a typewriter or a word processor that causes a tabulation 标签；选项卡

link [lɪŋk]	n.	sth which connects two other parts 联系
label ['leɪbl]	vt.	put or write a label on 贴标签于
accurately ['ækjərətli]	adv.	with few mistakes 准确地
familiar [fə'mɪliə(r)]	adj.	that you know well 熟悉的
promote [prə'məʊt]	vt.	help the progress of 促进
directory [də'rektəri]	n.	(computer science) a listing of the files stored in memory (usually on a hard disk) [计] 目录
award [ə'wɔːd]	n.	a prize such as money, etc. for sth that sb has done 奖；奖品
contact ['kɒntækt]	n.	the state of having a connection or exchanging information or ideas with sb else 联系；联络
layout ['leɪaʊt]	n.	arrangement 布局；安排；版面设计
consistent [kən'sɪstənt]	adj.	in agreement or accordance 一致的；符合的
animation [ˌænɪ'meɪʃn]	n.	the making of animated cartoons 动画片
bandwidth ['bændwɪdθ]	n.	the maximum amount of information (bits/second) that can be transmitted along a channel 通带宽度
frustrated [frʌ'streɪtɪd]	adj.	disappointingly unsuccessful 失意的；沮丧的
search engine		搜索引擎
as well as		也；还；并且
in case		万一；以防
eat up		吃光

Find Information

Task 1 Read the passage carefully and then answer the following questions.

1. What is the passage about?
___

2. How do you made your website attractive to visitors?
___

3. What should you pay attention to when you use links?
___

4. What is the most common way to promote a website?
___

5. Why can't you use too many graphics and animations?
___

Task II Read the passage carefully and then decide whether the following statements are true (T) or false (F).

( ) 1. We needn't think about what the visitors are looking for when we design a website.
( ) 2. Proper font and color should be chosen when designing a website.
( ) 3. We should put contact information on the website.
( ) 4. The layout and design should be consistent.
( ) 5. It's OK for the page to display in five minutes.

Words Building

Task I Translate the following phrases into English.

1. 网站设计_____
2. 字体和颜色的选择_____
3. 文本链接_____
4. 主页_____
5. 搜索引擎_____
6. 有奖站点_____
7. 联系信息_____
8. 网页布局_____

Task II Translate the following sentences into Chinese or English.

1. Make your site attractive to them by adding information that they are looking for.

2. The website can be made worse by bad font and color choices for your text.

3. Make sure that all your pages have a link back to your main page.

4. The most common way to promote a website is through search engines, directories, award sites, emails as well as links from other websites.

5. Avoid using too many graphics and animations as these files eat up bandwidth with your server.

6. 在设计您的网站时，应该遵循一些基本规则。

7. 您的网站应该容易操作。

8. 网站上应有你的联系信息以防有人需要联系你。

9. 保持你的网页布局和设计一致和连贯。
_____

10. 您的网页应该在 10 秒内打开。
_____

Task III Choose the best answer from the four choices A, B, C and D.

( ) 1. When _____ with people, you must look at their eyes.
  A. speaks   B. speak   C. to speak   D. speaking

( ) 2. He is always made _____ late without overtime pay.
  A. work   B. to work   C. working   D. works

( ) 3. Either I or he _____ going to design the website for you.
  A. are   B. am   C. is   D. will

( ) 4. He grows vegetables _____ flowers in his free time.
  A. as well as   B. too   C. also   D. as well

( ) 5. Take an umbrella _____ it rains.
  A. unless   B. in case   C. until   D. although

Task IV Fill in the blanks with the proper form of the word given.

1. Cheaper cars will _____ more consumers. (attractive)
2. _____ is difficult on this river because of the hide rock. (navigate)
3. Hearsay definitely can't be regarded as _____ information. (accurately)
4. The old man lives _____ with the laws of health. (consistent)
5. They made their preparations to _____ the conspiracy. (frustrated)

Cheer up Your Ears

Task I Listen and write down what you've heard. Then read and recite till you can use them fluently.

1. When designing your website, you should follow some _____ .
2. Make your site _____ to them by adding information that they are looking for.
3. The website can be made _____ by bad font and color choices for your text.
4. Your website should be easy to _____ .
5. Whichever you use, make sure they are labeled clearly and _____ .
6. Make sure that all your pages have a link back to your _____ .
7. Put your contact information on your website _____ someone needs to contact you.
8. Keep your web page layout and design _____ .
9. Your website should be quick to _____ .
10. Avoid using too many graphics and animations as these files _____ bandwidth with your

server.

Task II Listen to 5 short dialogues and choose the best answer.

1. A. On the second floor.　　B. At a men's store.
   C. In the women's department.　　D. In a department store.
2. A. Not everyone in England likes to read all the time.
   B. People who teach English may like other things besides books.
   C. The English like to read a lot and listen to music.
   D. An English teacher usually likes to read a lot.
3. A. It is not very good.
   B. It'll be continued next week.
   C. The woman probably won't attend it.
   D. The lecture will be busy next week.
4. A. What the whole internet community does.
   B. What "going online" means.
   C. How to surf the web.
   D. How to send e-mail.
5. A. A new application for the internet.
   B. How the internet meets all your needs.
   C. How the internet is great at offering particular information.
   D. How the internet has so much information that can tell you everything.

Table Talk

Group Work

Talk about the website design tips with your partner as much as you know.

_____

_____

Section Four Translation Skills

科技英语的翻译方法与技巧——词汇的减译和转译

英、汉两种语言，由于表达方式不尽相同，将英语翻译成汉语时有时需要减词，有时需要增词，有时需要转换词性或句子成分，这样才能符合汉语的表达习惯。同时在用英语表达中文意思时，也要注意这种翻译特点，养成英语思维的习惯，这样才能说出地道的英语，切忌将英语的某词与中文的某词机械地一一对应。

1. 减译

在进行翻译时，原文中的某些词可不必翻译出来，这就是通常说的省译或减译。在不影响意思表达的基础上，一些形式主语、连词、冠词、介词等虚词，一些可有可无或者有了反而觉得累赘的词都可以删减，以符合汉语表达习惯。例如：

It is important to recognize that, in a turning operation, each cutting pass removes twice the amount of metal indicated by the cross slide feed divisions.

在外圆车削过程中，每次进刀的金属去除量是横向拖板进给刻度指示的两倍，认识到这一点很重要。it, that 这些虚词可以不译。

These rigid machine tools remove material from a rotating workpiece via the linear movements of various cutting tools, such as tool bits and drill bits.

这些坚硬的机床通过各种刀具（如车刀和钻头）相对旋转工件作直线运动从而去除材料。这里并 bits 表示带刀尖的刀具，车刀和钻头就是这样的刀具，因此可不译。

You can use the carriage hand wheel to crank the carriage back to the starting point by hand.

可以使用横拖扳手轮将横托板摇回起点。这里 crank 就有用手摇的意思，因此 by hand 显得有点累赘，可以省译。

（1）介词的减译

英语中介词丰富，通常用来表示名词、代词等和句中其他词之间的关系。汉语中也有介词，但数量有限，词与词之间的关系主要通过词序和意会来区分。因此，在英译汉时减译的情况较多。

The temperature of metal moulds should always be in excess of 200℃.

金属模的温度应该经常保持在 200℃以上。

（2）冠词的减译

英语冠词是附在普通名词前的虚词，是使用频率最高的功能词之一，但其本身没有意义，只对名词起限定和辅助说明的作用。汉语中没有冠词，在将英语译成汉语时，除少数情况外，冠词一般减译。例如：

A liquid metal becomes a gas at or above its boiling point.

在沸点或沸点以上，金属液会变成气体。

另外，用在固定词组和固定搭配中的冠词一般都应减译。例如：

as a consequence	因此，从而
in a few words	简言之
on an average	平均来说
in the end	最后，终于
make the grade	达到标准，合乎要求
hit the mark	中标，达到目的

冠词并非在任何情况下都可减译。当不定冠词具有明显的数字概念或与单数可数名词连用，表示单位数量时，必须译出；而定冠词在起着特指某具体事物的作用时，通常也减译。

（3）动词的减译

英语和汉语的动词都用来表示动作或存在的状态，但英语动词的使用范围远远超过汉语动词，英语的每个句子都少不了一个动词，而且句子的谓语总要由动词充当。而汉语句子的谓语除动词外还可以直接用名词、形容词或词组来担任。在英译汉时可视具体情况把原文的某些动词减译。例如：

The following provides a brief review of applicable theory for pressure casting.

下面简要评述一下适用于压力铸造的理论。

（4）代词的减译

①人称代词的减译。汉语中，如果紧连的两个句子主语相同，其中一句就可以不用主语。而英语则每句必有主语，常在一句中用人称代词表示和前句相同的主语。这样使用的英语人称代词在英译汉时应尽量减译。

②物主代词的减译。英语中的物主代词一般不能省略，而汉语中在关系明确的情况下不必再用物主代词。英译汉时减译英语物主代词的情况十分常见。例如：

One cannot see the metallographic structure of alloy with one's naked eyes.

人们不能用肉眼看到合金的合金组织。

（5）根据修辞需要进行的减译

在专业英语中，常有对同一概念用两个词重复表达的现象，这两个词中前一个往往是专业词汇，后一个往往是普通词汇。两者在语法上是同位关系，这样做是为了适应不同层次读者的需要。但在译成汉语时，为了使概念准确，只需译出其中一个汉语中通用的名称。例如：

The advantage of induction furnace melting is that it melts the metal with minimum element loss.

感应炉熔炼的优点是元素损失很少。

（6）引导词的减译

英语的副词 there 可以和动词 to be 及某些不及物动词连用，表示"存在，有"。这一句型中的 there 已经变成英语中特有的引导词，不再具有副词"在那里"的含义。汉语中没有这种形式和结构，因此应减译。英语代词 it 已不再具有代词"它"的意义，汉语中也没有相应的结构，所以英译汉时也应减译。例如：

If more thought were given to the consequences of injuries, there would be fewer tendencies to ignore safety precautions and thus, fewer accidents.

如果更多地考虑伤害后果，忽视安全防范的倾向就会减少，从而减少事故。

2. 转译

由于英汉两种语言表达方式不同，语言形式差别很大，为了使译文符合汉语表达习惯，除了运用词类转译技巧之外，往往还伴随着句子成分的转换。例如，在某些特定的句型中，英语句子的状语、定语、表语可能转译成汉语句子的谓语等。下面仅列举常见的几种情况，供举一反三、类比推敲。

（1）词类转译

①动词转换成名词

It is important to recognize that, in a turning operation, each cutting pass removes twice the amount of metal indicated by the cross slide feed divisions.

在外圆车削过程中，每次进刀的金属去除量是横向托板进给刻度指示的两倍，认识到这一点很重要。

Rooms with constant temperature and clean environment are used in assembling and inspecting precision parts such as spindles.

恒温超净室用于机床主轴部件等精密部件的装配及检验。

②名词转译成汉语动词

If you sense an impending collision, press the EMERGENCY STOP button.

如果有可能碰撞，按下 EMERGENCY STOP（紧急停止）按钮。

（2）句子成分的转换

①转换成汉语主语。在英语"there be"句型中，往往将状语省略，去掉介词并转换成汉语主语。例如：

LGMazak equips a MAZATROL system and FMS (Flexible Manufacturing System) on every machine, and connects them on the local network.

LGMazak 的每一台机床都配置了 MAZATROL 系列数控系统和柔性制造系统，并将这些机床连接到公司内的局域网。

There are a variety of machine tools in the workshop.

这个车间有各种各样的机床。

英语中介词短语作定语时，有时可以转换成汉语主语，同时略去介词不译。

②介词短语转换为定语

The following is a list of required regular maintenance for a horizontal machining center.

以下是卧式加工中心的常规维护事项。

These workers first review blueprints or written specifications for a job.

首先，这些工人阅读作业零件的图纸或书面说明。

③转换为汉语状语从句

There is a lot of manual intervention required to use a drill press to drill holes.

用台式钻床钻孔时，需要很多人工干预。

Listening

Task 1 Listen to 2 conversations and choose the best answer.

Conversation 1

1. A. One year.　　　　　　　　B. Five years.
　C. Three years.　　　　　　　D. Seven years.

2. A. Because he expects a better salary.
　B. Because he is tired of his boss.
　C. Because he doesn't like traveling.
　D. Because he likes to work in a big company.

3. A. In three working days.　　　B. the next day.
　C. Within two weeks　　　　　D. A month later

Conversation 2

4. A. To the Science Museum.　　B. To the History Museum.
　C. To the Art Museum.　　　　D. To the Space Museum.

5. A. About every four minutes.　　B. About every five minutes.
　C. About every six minutes.　　　D. About every seven minutes.

Unit Seven Website

Task II **Listen to the passage and fill in the blanks with the missing words or phrases.**

Welcome to this electronics exhibition. Here you can find almost all of the ___1___ products of our company. Let's go to the ___2___ section first. Please follow me. The products on display in this section are especially designed for ___3___ who wish to use a computer at home. Look at the keyboard. Its big keys are particularly easy to use by whose ___4___ is not good enough. Here, please take a close look at the mouse. It can fit all size of hands. Besides, it has a special sound effect. Listen, not only are the sounds pleasant, but the two buttons ___5___ different sounds. Now I'll give you three minutes to try by yourselves. Then we'll go on to the software section.

Extensive Reading

Directions: After reading the passage, you are required to complete the outline below briefly.

As you're designing your new website, you'll be tempted with web design ideas that could turn into fatal mistakes. There are five of the most common mistakes to avoid at all costs below.

1. Too Many Graphics

Having too many graphics can cause your site to load entirely too slow. Visitors will get impatient and often click out of your site — never return.

2. Counters

A visitor counter or hits counter should not be seen on your site unless you have a tremendous number of visitors. The only way showing a counter is advantageous is that you've had millions of visitors and wish to display the popularity of your site or would like to attract advertisers with the large number.

3. Banners

Limit your banners to the bare necessities. Why? Because banners are graphics that can slow loading time and are a turn-off for many surfers on the internet. For most, "banner" is just another word for "ad" and they avoid clicking on them.

4. Scattered website

Organizing your site to lead visitors is very important whether you're leading them to buy something or just to click and go to another place in your site.

5. Generalization

The most effective way of selling on the Internet is to personalize your website to reach your target audience.

Five Most Common Web Design Mistakes

These five mistakes should be avoided at all costs if you want to build an effective and successful online business.

1. Too many graphics can cause your site to load too ___1___ . Visitors will get ___2___ .
2. A counter shouldn't be seen unless you have a ___3___ number of visitors and you wish to

display the ___4___ of your site.
3. Limit your banners to the ___5___ because they can slow loading time and surfers avoid ___6___ on them.
4. Organizing your site to ___7___ is very important.
5. Your website must be personalized to reach your ___8___.

Section Five Writing

Fax 传真

所谓传真通信，是把记录在纸上的文字、图表、相片等静止的图像变换成电信号，通过连接电话线的传真机来发送，在接收方获得与发送原稿相似的记录图像的通信方式。传真文稿的格式简单、操作方便，可以按照信函的格式书写。

英语传真格式：

传真既有信函的特征又有备忘录、电子邮件的特点，业务传真构成如下。

1. 寄件人公司名称和地址：一般公司都有固定的传真格式，也称传真头，包括公司名称、地址、电话、传真和电子邮箱。
2. 寄件人
3. 文件编号：为了方便存档和日后查阅，公司将寄发或收到的传真、信函或者其他一些资料文件，加以编号后形成的索引码，也叫"档号"。由于双方公司都会给各自的文件编号，希望对方在回复时指明是针对哪份文件作出的答复，所以就会有 your ref 和 our ref。
4. 日期
5. 收件人公司名称
6. 收件人姓名
7. 收件人传真号码
8. 收件人电话
9. 传真页码
10. 主题（事由）
11. 传真电文：包括称呼、正文、结束语和签名。

书写传真要简明、清楚、正确无误。

英语传真常用语：

1. To	收件人或单位
2. Attn (Attention)	收件人
3. Fax	传真号码
4. Tel	电话号码
5. From	寄件人
6. Pages	传真页码
7. Date	日期
8. Ref (reference)	编号
Our Ref	我方编号，即发信人编号
Your Ref	贵方编号，即收信人编号

9. Re / Subject　　　　　　　　　　　　主题（事由）
10. I look forward to your prompt reply.　　期待回复。
11. Best regards.　　　　　　　　　　　　祝好！

Read and understand the sample fax.

Sample

To: Columbia University　　　　　　From: Mr. Wang Qian
Attn: Mr. George Smith　　　　　　　Ref No: FZ856
Fax: 01-2128545333　　　　　　　　Date: March 3rd, 2016
Tel: 01-2128542773
Pages: 1
Subject: Invitation
Content:
Dear Mr. Smith,
　　We'll be holding a conference on learning as a Second Language in Beijing, China on May 15th, 2015. You are welcome to join the conference if you're interested in it. Please contact us before April 15th so that we can prepare a formal invitation letter for you.

　　　　　　　　　　　　　　　　　　　　　　　　　　　　Best regards,
　　　　　　　　　　　　　　　　　　　　　　　　　　　　Wang Qian

范例

北京外国语大学
北京市海淀区西三环北路 19 号（邮政编码：100089）
电话：86-01-68422277　　　传真：86-01-68423144
电子邮箱：bwxzb@mail.bfsu.edu.cn

致：哥伦比亚大学　　　　　　　寄件人：王谦先生
收件人：乔治·史密斯先生　　　　文件编号：FZ856
传真：01-2128545333　　　　　　日期：2016 年 3 月 3 日
电话：01-2128542773
页数：1
主题：邀请
内容：
亲爱的史密斯先生，
　　有关第二语言——英语学习大会将于 2015 年 5 月 15 日在中国北京举行。我们真诚邀请您参加。请于 4 月 15 日前告知我们您是否接受我们的邀请以便给您寄发正式邀请函。
　　敬候回复

　　　　　　　　　　　　　　　　　　　　　　　　　　　　祝好！
　　　　　　　　　　　　　　　　　　　　　　　　　　　　王谦

Practice

You are required to fill in a Fax according to the following information in Chinese.

要点：下面是 2015 年 10 月 10 日发出的一份传真，内容有关丁晓到达的日期。

收件人：丁晓　　　传真：0411-76213453　　　寄件人：Peter Green

Griffith University
Faculty of Education and Arts
Gold Coast Campus Parklands Drive Southport
Telephone: 076948803　　　Fax: 076948534
E-mail: petergreen@eda.gu.edu.au

To: _____
（北京大学英语系）_____
Fax: _____
From: _____
Date: _____
Pages: 1
Re: _____
（亲爱的丁晓：）_____:
（谢谢你 2015 年 9 月 12 日的来信。我理解你的问题。你可以 12 月来此。请随时告知。）

_____
_____.
　　　　　　　　　（祝好。）_____
　　　　　　　　　（彼特·格林）_____

Section Six　Grammar

Subjunctive Mood II 虚拟语气（二）

虚拟语气表示说的话不是事实，而是与事实相反的假设，也可是一种愿望、建议、要求等。常用于表达不可能实现的假设或愿望的虚拟方式为时态变化型。

今天，让我们学习另一种常见的虚拟方式——should 型虚拟。

请牢记从句固定结构为：主语 +（should 可省）+ 动词原形。

从句虚拟时，不必考虑时态和单复数，但必须区分语态。

☆ 主动语态的从句结构为：主语 +（should 可省）+ do

☆ 被动语态的从句结构为：主语 +（should 可省）+ be done

常用 should 型来表达虚拟语气的情况有：

1. 表建议、命令、要求的词，（词性不限）后接从句时，用 should 型虚拟。如：

（建议）　　　　suggest, advise, propose, recommend
（命令）　　　　order, command
（要求）　　　　require, request, demand, ask, desire
（坚持要求）　　insist

例：（1）有人建议她坐公汽去大连。

Someone suggested (that) she（should）go to Dalian by bus. 动词后宾语从句虚拟 =There was a suggestion that she go to Dalian by bus. 名词后同位语从句虚拟 = It was suggested that she go to Dalian by bus. 形式主语 It+ 分词的主语从句虚拟

（2）老师要求我们今天必须完成作文。

The teacher demanded (that) we (should) finish our composition today. 从句用主动语态 = The teacher demanded our compositions（should）be finished today. 从句用被动语态

（3）医生建议他不要去那里。

The doctor advised (that) he not go there. 宾语从句虚拟 = The doctor's advice was that he (should) not go there. 表语从句虚拟

（4）He desires that he (should) be allowed to go swimming. 他要求被允许去游泳。

（5）Our only request is that it (should) be settled as soon as possible.
我们唯一的请求就是尽快解决这个问题。

（6）The workers' requirement was that their working conditions be improved.
工人们的要求是改善工作条件。

注意：suggest 作"建议"解时，从句虚拟，但意为"表明、暗示"时，从句不虚拟。

例：When the son suggested that they should go to the park on Sunday, the expression on his mother's face suggested that she agreed with him.

当儿子提议星期天去公园时，母亲的表情表明了赞同。

只有 insist 意为"坚持要求，强调动作"时，从句用虚拟语气，

但意为"坚持说、坚持认为"时，从句不用虚拟语气。

例：The boy insisted that he wasn't wrong. 这个孩子坚持说他没有错。

2. It 固定句型中的主语从句，虚拟方式用 should 型。

常见结构为：It is /was + 形容词 / 名词 / 分词 + that 引导的主语从句虚拟

（主语 +should+ 动词原形）

注意：should 在此是助动词，本身无实义（可省），从句中 should 有时带有感情色彩。

例：It's strange that she should marry such an ugly man. 她竟然会嫁给那样一个丑陋的男人，真奇怪！

这些形容词主要表示必要性、重要性、强制性、义务性等，常见词有：

（必要的）　　　necessary, imperative, essential
（重要的）　　　important, crucial, urgent（紧急的）, natural（自然的）
（奇怪的）　　　strange, impossible, incredible（难以置信的）

常见名词有：pity（遗憾），duty（义务、责任）

常见分词有：arranged（安排），proposed（建议），recommended（推荐）

例：

（1）我们必须保护野生动物。

It's necessary that we should protect wild animals.
= It's necessary that wild animals should be protected.

（2）It is important that university graduates should have not only theory but also practice.
大学毕业生不仅要有理论知识，更要有实践经验，这一点很重要。

（3）It is natural that a person with poor eyesight should wear glasses or contact lens.
近视的人配戴眼镜或是隐形眼镜是很自然的事情。

（4）It is arranged that he should be sent to Beijing right away.
已安排将他立即派往北京。

3. lest（以免；免得）/ for fear that（生怕，唯恐）/ in case（以防）后接目的状语从句表消极意义时，虚拟方式用 should 型。例如：

（1）You must wake him up early lest he (should) be late for school.
你务必早点叫醒他，以免他上学迟到。

（2）She is now studying hard for fear that she (should) fail in the English test.
她现在很勤奋，生怕英语不及格。

（3）Keep quiet in case you (should) interrupt him when he is busy working.
要保持安静，以防在他忙于工作时打扰到他。

但 in case 从句还可以是陈述语气。例如：
Keep the window closed in case it rains. 把窗户关好，以防下雨。

此外，在 It's time 句型中，其后定语从句要虚拟，且有两种虚拟方式。
其含义是"该做某事的时候了"
"It is (high/about) time +(that) 从句"句型中，that 常常省略。
从句虚拟方式常用的一种为将从句的动词用一般过去时 (did) 表达，有时用另一种：主语 +should（不可省略）+ 动词原形。例如：

① It is time the children went to bed. 孩子们早该上床睡觉了。
② It is time we left. 我们该走了。

☆ 最后，了解一下特殊的虚拟语气用法：
A. 表示"祝愿"句型中的虚拟语气。（常用动词原形或 may 引导的倒装句。）例如：
① Long live the Communist Party of China! 中国共产党万岁！
② God bless you! 愿上帝保佑你。(或说成 May God bless you)

B. 虚拟过去除使用 would have done 过去本会去做却没做……之外，
还可用情态动词 should / could + have done 表示跟过去事实相反的情况。
请牢记：should have done 过去本该做却没做……= ought to have done
could have done 过去本可以做却没做…… 例如：

① Tom, you are too lazy. The work should have been finished yesterday.
　 汤姆，你太懒惰了。这项工作本来应该昨天就做完成的。

② He could have passed the exam, but he was too careless.
　 本来他能够通过考试，但是他太粗心。

此外，还要学会区分并牢固掌握以下结构的含义和用法。
助动词 may/might / must /can't + have done 表示猜测过去的事情。
may/ might have done 过去可能做了……

must have done   过去肯定做了……
can't have done   过去不可能做了……
例：He might have said it yesterday. 他昨天可能说了。
He must have been here yesterday. 他昨天一定到过这儿了。
Mr. Smith can't have gone to Beijing, for I saw him in the library just now.
史密斯先生不可能去北京了，我刚才还在图书馆见过他。

Exercises

I. Fill in the blanks with the proper form of the words given in brackets.

1. Mary insisted that Tom _____ her money back at once. (give)
2. My suggestion was that the meeting _____ off till next week. (put)
3. The boss's order is that everyone _____ their own work in time. (finish)
4. It is necessary that the problem _____ at once. (solve)
5. I demanded that I _____ to call my lawyer. (allow)
6. It's time we _____ . If we don't hurry, we'll be late. (leave)
7. It's high time you _____ your schoolwork more seriously. Graduation is right around the corner! (take)
8. It is important that he _____ a decision before Friday. (make)
9. The boss demanded that the work _____ before eleven. (finish)
10. He got up early that morning lest he _____ the train. (miss)

II. Choose the best answer.

(　) 1. It was suggested at the meeting that everyone _____ a report.
　　A. must make　　B. make　　C. will make　　D. should makes

(　) 2. Your advice that she _____ till next week is reasonable.
　　A. will wait　　B. is going to wait　　C. waits　　D. wait

(　) 3. My father did not go to New York. The doctor suggested that he _____ there.
　　A. not went　　B. won't go　　C. not go　　D. not to go

(　) 4. You look so tired tonight. It is time you _____ .
　　A. go to sleep　　B. going to sleep　　C. to go to bed　　D. went to bed

(　) 5. I could have done better if I _____ more time.
　　A. have had　　B. had　　C. had had　　D. will have had

(　) 6. The law requires that everyone _____ his car checked at least once a year.
　　A. has　　B. had　　C. have　　D. will have

(　) 7. It's high time _____ a decision.
　　A. I'm making　　B. I'll make　　C. I made　　D. I make

(　) 8. The librarian insists that John _____ no more books from the library before he returns all the books he has borrowed.
　　A. will take　　B. took　　C. take　　D. takes

(　) 9. Lisa's pale face suggested that she _____ ill and her parents suggested that she

_____ medicine.

A. be; should take B. was; take C. was; takes D. be; took

( ) 10. The job would require that _____ at 7 o'clock every morning.

A. he will be at the factory B. he be at the factory
C. he was at the factory D. he has been at the factory

III. Find and correct the mistakes.

1. I suggested that the meeting will be held at once.
2. It's important that we worked out a plan.
3. It's high time the government does something about the employment problem.
4. It is necessary that the sick boy should take to the nearest hospital at once.
5. He made a suggestion that Ann cleans the house twice a day.
6. His grandmother must be a beautiful lady forty years ago.

IV. Fill in the blanks and finish the dialogue.

Lucy: Dad! Tomorrow is my mother's fortieth birthday!

Dad: Yes. How time flies! I wish we ___1___ (be) twenty years younger.

John: Come on, Dad! You are still young. It's time we ___2___ (decide) where to celebrate our mother's birthday.

Lucy: I suggest we ___3___ (try) to cook Sichuan food at home because our mother likes spicy food.

Dad: John! Do you agree to Lucy's suggestion that we ___4___ (stay) at home?

John: No, I don't agree. My proposal is that we ___5___ (go) to the most popular Sichuan food restaurant in our city. There are all kinds of spicy food. We can order whatever we like.

Lucy: Sounds great!

Dad: I'm sure your mother will like it. But it's necessary that a table ___6___ (reserve) in advance.

John: Don't worry! I'll make a reservation right away.

CULTURE TIPS

扫一扫认识网站人员构成

Unit Eight
Search Engine

Section One Warming Up

Pair work

Talk about the following search engines. Say as much as you can.

Section Two Real World

Listen and act the following dialogue which is about how to choose a cell phone with the help of search engines between two students.

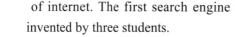

Candy: I want to buy a cell phone. How can I **choose** from such a great **variety**?

Rex: It's a piece of cake. Use the search engine and have a look at the **evaluation** and **recommendation** from the users.

Candy: Search engine? What do you mean?

Rex: Well, let me put it this way. According to your requirements, the search engine search for **relevant** information, then **organizes** and processes them and provides **retrieval** services. Finally it **presents** the results to the user.

Candy: Sounds a little **complicated**. You seem to know much about search engines. Do you know when the search engine was invented?

Rex: Yeah. It came out with the development of internet. The first search engine was invented by three students.

Candy: Really? That's **incredible**!

Rex: Surprised? In fact, many Internet **related technology** and **products** are invented by students because young people are in a more active state of mind.

Candy: We should work hard. Maybe we'll get something invented.

Words & Expressions

choose [tʃuːz]	vt. & vi.	to decide which person or thing that are available you want to have 挑选；选择
variety [vəˈraɪəti]	n.	a particular type that is different from others in a group to which it belongs 种类；品种
evaluation [ɪˌvæljuˈeɪʃn]	n.	an appraisal of the value of something 评价
recommendation [ˌrekəmenˈdeɪʃn]	n.	the act of recommending or sth. (especially a course of action) that is recommended 推荐
relevant [ˈreləvənt]	adj.	connected with the subjects 有关的
organize [ˈɔːgənaɪz]	vt. & vi.	form (parts) into a whole; put into working order 组织
retrieval [rɪˈtriːvl]	n.	(computer science) the operation of accessing

		information from the computer's memory [计]检索
present [prɪˈzent]	vt.	give or hand sth to sb 提交
complicated [ˈkɒmplɪkeɪtɪd]	adj.	difficult to understand or explain because there are many different parts 难懂的
incredible [ɪnˈkredəbl]	adj.	unbelievably good 难以置信的
related [rɪˈleɪtɪd]	adj.	being connected either logically or causally or by shared characteristics 相关的
technology [tekˈnɒlədʒi]	n.	knowledge dealing with scientific and industrial methods and their practical use in industry 技术
product [ˈprɒdʌkt]	n.	thing or substance produced by a natural or manufacturing process 产品；产物
cell phone		手机
a piece of cake		小菜一碟
search engine		搜索引擎
have a look at		看一看
according to		根据
search for		搜索
come out		出现
in an active state of mind		思维活跃

Find Information

Task I Read the dialogue carefully and then answer the following questions.

1. What does Candy want to buy?

   _____

2. How does Rex advise Candy to choose?

   _____

3. When was the first search engine invented?

   _____

4. Does Rex think young people are in a more active state of mind?

   _____

5. What do they do?

   _____

Task II Read the dialogue carefully and then decide whether the following statements are true (T) or false (F).

(   ) 1. Rex thinks it easy to choose a cell phone from such a great variety.

(　) 2. The search engine can presents the results according to the user's requirements.
(　) 3. The first engine was invented by four students.
(　) 4. The first engine was invented in the nineteenth century.
(　) 5. Candy is surprised at what Rex said.

Words Building

Task I Choose the best answer from the four choices A, B, C and D.

(　) 1. This is _____ funny story that we all enjoy listening to it.
　　　 A. so　　　　　　B. such a　　　　　C. such　　　　　D. so a
(　) 2. Missing the train means _____ for another hour.
　　　 A. to wait　　　　B. waits　　　　　 C. wait　　　　　D. waiting
(　) 3. This magazine _____ once a month.
　　　 A. comes out　　　B. comes across　　C. comes on　　　D. comes up
(　) 4. _____, what is your next goal ?
　　　 A. In the way　　 B. On the way　　　C. By the way　　D. In this way
(　) 5. In cold weather, it's hard to get the car _____ .
　　　 A. starts　　　　 B. started　　　　 C. to start　　　D. starting

Task II Fill in each blank with the proper form of the word given.

1. The school has only been open for six months, so it's too early to _____ its success. (evaluation)
2. I was asked to _____ the trip, but I messed it up. (organization)
3. There is no need to _____ matters. (complicated)
4. The government has done a lot to promote _____ development. (technology)
5. The new play is _____ boring. (incredible)

Task III Match the following English terms with the equivalent Chinese.

A — dynamic page　　　　　　　　1. (　) 搜索引擎排名
B — algorithm　　　　　　　　　　2. (　) 关键词密度
C — repeated word　　　　　　　　3. (　) 按点击数付费
D — hidden text　　　　　　　　　4. (　) 垃圾邮件
E — spam　　　　　　　　　　　　5. (　) 动态网页
F — meta refresh　　　　　　　　 6. (　) 刷新标识
G — search engine ranking　　　　7. (　) 蜘蛛
H — spider　　　　　　　　　　　 8. (　) 算法
I — keyword density　　　　　　　9. (　) 重复性关键词
J — cost per click　　　　　　　 10. (　) 隐藏文本内容

Cheer up Your Ears

Task I Listen and write down what you've heard. Then read and recite till you can use them fluently.

1. How can I choose from such a great _____ of cell phones?
2. Use the search engine and _____ at the evaluation and recommendation from the users.
3. According to your requirements, the search engine _____ relevant information.
4. The search engine _____ with the development of internet.
5. Search engine optimization is an ongoing process that can _____ .
6. Search engines are programmed to _____ websites based on some combination of their popularity and relevancy.
7. The search engines _____ every time someone clicks on one of these ads.
8. Some search engines _____ an advanced feature called proximity search.

Task II Listen and fill in the blanks with the missing words and role-play the conversation with your partner.

A: There are __1__ web pages on the net. How do I find what I'm looking for?
B: Use a __2__ like Google.
A: How does it work?
B: You enter the name or topic you are __3__ in then ask the search engine to find pages about your topic.
A: It must take a long time to search all those pages.
B: Not really, usually __4__ a minute.
A: What search engine do you usually use?
B: I usually use several different search engines __5__ . Different sites are registered with different search engines, and different engines use different logic to search. So they will all turn up different results.

Task III Listen to 5 recorded questions and choose the best answer.

1. A. He is leaving by bus.　　　　B. He has a big family.
   C. He is living in a small town.　D. He is working as a lawyer.
2. A. Sure. Here you are.　　　　 B. Yes, please give it to me.
   C. Sorry, I can't help you.　　　D. No, I can take it myself.
3. A. Yes, you can drive it.　　　　B. Should I go with you?
   C. No, everything is fine.　　　 D. Take your time. There is no hurry.
4. A. Yes, I got it.　　　　　　　　B. It's interesting.
   C. No, I don't have it.　　　　　D. It begins at 6:00.
5. A. It's difficult to park here.　　 B. I don't like to take a taxi.

C. I came here by bus.   D. I agree with you.

 Table Talk

Task I Complete the following dialogue by translating Chinese into English orally.

A: Human flesh search is popular now. What do you ___1___ （认为）it?

B: ___2___ （依我看）it becomes an efficient tool to find desired information. Those immoral people have been made public.

A: Yeah, they would get ___3___ （至少）moral punishments. In this sense, it proves to be an effective means to find out norm violators and even criminals and give them a dose of their own medicine.

B: But human flesh search is accused of intruding ___4___ （个人隐私）and in some cities, authorities have tried to prohibit the search.

A: Privacy is the basic right for everyone, even for the criminal.

B: ___5___ （此外）, human flesh search can be misused on purpose. If you don't like someone, you can lie about him. So strangers will attack him, which will bring a lot of troubles.

A: I believe that, with the gradual perfection of human-powered search, it will receive a wide application.

Task II Pair-work. Role-play a conversation about search engines with your partner.

Situation

Imagine you want to look for some information with the help of search engines. Now you are talking with your friend.

Section Three Brighten Your Eyes

Pre-reading questions:

1. When you want some information, who do you ask for help?
2. Which search engine is the most popular in China?

How Web Search Engines Work

Web search engines work by **storing** information about many web pages, which they **retrieve** from the HTML of the pages. These pages are retrieved by a web **crawler** (sometimes also known as a spider).

The search engine then **analyzes** the contents of each page to **determine** how it should be **indexed** (for example, words can be **extracted** from the titles, page content, or special fields called meta tags). **Data** about web pages is stored in an index **database** for use in later **queries**. A query from a user can be a single word. The index helps find information relating to the query as quickly as possible.

When a user enters a query into a search engine, the engine examines its index and provides a listing of best-matching web pages according to its **criteria**, usually with the document's title and sometimes parts of the text.

The usefulness of a search engine depends on the **relevance** of the result set it gives back. While there may be millions of web pages that include a **particular** word or phrase, some pages may be more **relevant**, popular, or **authoritative** than others. Most search engines employ methods to **rank** the results to provide the "best" results first.

Words & Expressions

store [stɔː(r)]	vt. & vi.	make up and keep a supply of or put away for future use 储藏；存放
retrieve [rɪˈtriːv]	vt.	find again or extract 检索
crawler [ˈkrɔːlə(r)]	n.	terrestrial worm that burrows into and helps aerate soil 爬虫
analyze [ˈænəlaɪz]	vt.	make a mathematical, chemical, or grammatical analysis of; break down into components or essential features 分析
determine [dɪˈtɜːmɪn]	vt.	fix or find out exactly 确定
index [ˈɪndeks]	vt. & n.	list in an index 指出；索引
extract [ˈekstrækt]	vt.	remove with a machine or instrument or by chemical means; calculate the root of a number 提取；榨出；[计算机] 提取
data [ˈdeɪtə]	n.	information that is stored by a computer 数据资料
database [ˈdeɪtəbeɪs]	n.	an organized set of data that is stored in a computer and can be looked at and used in various ways（储存在计算机中的）数据库；资料库
query [ˈkwɪəri]	n.	an instance of questioning 疑问；质问
criteria [kraɪˈtɪəriə]	n.	an established rule, standard, or principle, on which a judgment is based 标准；准则（名词 criterion 的复数形式）
relevance [ˈreləvəns]	n.	the relation of something to the matter at hand 关联；相关性
particular [pəˈtɪkjələ(r)]	adj.	one only, and not any other 特定的；special or more than usual 特殊的
relevant [ˈreləvənt]	adj.	connected with the subjects 有关的；切题的

authoritative [ɔːˈθɒrətətɪv]	adj.	generally regarded as providing knowledge or information that can be trusted 权威性的
rank [ræŋk]	vt.	take or have a position relative to others 分等级；排名
meta tags		元标签
relating to		关于；与……有关
as ... as possible		尽可能……
according to		根据；依照
depend on		依靠；取决于
millions of		数以百万计的

Find Information

Task I Read the passage carefully and then answer the following questions.

1. What is the passage about?

2. By what are the web pages retrieved?

3. Where is the data about web pages stored?

4. What is provided when a user enters a query into a search engine?

5. What does the usefulness of a search engine depend on?

Task II Read the passage carefully and then decide whether the following statements are true (T) or false (F).

( ) 1. Web search engines store information about many web pages which are retrieved by a spider.
( ) 2. A query can only be a word.
( ) 3. The search engine provides a listing of best-matching we pages with the whole content of the document.
( ) 4. The usefulness of a search engine depends on how much web pages it indexes.
( ) 5. Most search engines provide the best results by employing methods.

Words Building

Task I Translate the following phrases into English.

1. 搜索引擎

2. 网络爬虫_____
3. 元标记_____
4. 输入查询_____
5. 最佳匹配的网页清单_____
6. 文件的标题_____
7. 结果集合_____
8. 排序结果_____

Task II Translate the following sentences into Chinese or English.

1. Web search engines work by storing information about many web pages, which they retrieve from the HTML of the pages.
   _____

2. The search engine then analyzes the contents of each page to determine how it should be indexed.
   _____

3. Data about web pages are stored in an index database for use in later queries.
   _____

4. When a user enters a query into a search engine, the engine examines its index.
   _____

5. While there may be millions of web pages that include a particular word or phrase, some pages may be more relevant, popular, or authoritative than others.
   _____

6. 这些网页由网络爬虫检索。
   _____

7. 可以从标题，网页内容，或叫做元标记的特殊领域提取单词。
   _____

8. 检索有助于尽快找到与查询相关的信息。
   _____

9. 一个搜索引擎的有效性取决于其反馈结果的相关性。
   _____

10. 大多数搜索引擎采用某些方法进行排序来提供"最好"的结果。
    _____

Task III Choose the best answer from the four choices A, B, C and D.

(　) 1. It's a good way to study English by _____ English movies.
    A. watch   B. to watch   C. watching   D. watched

(　) 2. I need your advice. Please write to me as _____ as possible.
    A. fast   B. quickly   C. soon   D. many

(　) 3. _____ people died in the war.
    A. Millions of   B. Two millions   C. Two millions of   D. Million of

( ) 4. He is _____ than the other singers in this city.

  A. popular   B. more popular   C. popularer   D. as popular

( ) 5. More measures _____ here to prevent the haze.

  A. take   B. have taken   C. have been taken   D. is taking

Task IV Fill in the blanks with the proper form of the word given.

1. The ship was buried, beyond _____ at the bottom of the sea. (retrieve)
2. The _____ of a name for the club took a very long time. (determine)
3. In what way does it _____ to your current career? (relating)
4. The baby is just learning to _____ . (crawler)
5. Have you heard of the _____ report? (relevance)

Cheer up Your Ears

Task I Listen and write down what you've heard. Then read and recite till you can use them fluently.

1. Web search engines work by _____ information about many web pages.
2. These pages are retrieved by a Web _____ .
3. The search engine then _____ the contents of each page.
4. Data about web pages are stored in an index _____ for use in later queries.
5. The index helps find information _____ the query as quickly as possible.
6. When a user enters a _____ into a search engine, the engine examines its index.
7. Then the search engine provides a listing of _____ web pages according to its criteria.
8. The usefulness of a search engine _____ the relevance of the result set it gives back.
9. Some pages may be more _____ , popular, or authoritative than others.
10. Most search engines employ methods to _____ the results to provide the "best" results first.

Task II Listen to 5 short dialogues and choose the best answer.

1. A. A fax.   B. A report.
  C. A newspaper.   D. A letter.
2. A. Buy a new computer.
  B. Restart the computer.
  C. Ask someone to repair the computer.
  D. Borrow a computer from the company.
3. A. She hasn't sent the email.   B. She hasn't got any e-mail.
  C. She won't read the email.   D. She won't reply to the e-mail.
4. A. The price of the books.   B. The author of the books.

C. The way to pack the books.   D. The time to get the books.
5. A. Very nice.   B. Very strict.
   C. Very humorous.   D. Very shy.

Table Talk

Group Work

Talk about the search engine with your partner as much as you know.

_____

_____

Section Four Translation Skills

科技英语的翻译方法与技巧——定语的译法

科技英语中，经常需要对某个概念进行定义，这时常使用定语或定语从句。翻译成汉语时，可将长的定语从句分为并列的两句或多个单句，对于短的定语则用"……的……"。

1. 过去分词作定语

过去分词作定语，通常句意多为被动，后置时相当于定语从句，只是省略了 which。例如：

The ways are the ground surfaces on the top side of the bed on which the carriage and tailstock ride.

导轨是床身上磨削的表面，床鞍和尾架跨骑在床身上。这里 ground 是过去分词作定语，已形容词化。又如，given cutting speed（给定的切削速度），desired position（期望的位置），tapered hole（锥孔）。

2. 现在分词做定语

The primary purpose of the tailstock is to hold the dead center to support one end of the work being machined between centers.

尾架的主要作用是安装死顶尖以支撑两顶尖之间的被加工工件的一端。

The ratio of the threads per inch of the thread being cut and the threads per inch of the lead screw is the same as the ratio of the speeds of the spindle and the lead screw.

被切削螺纹的每英寸螺纹大小和丝杠的每英寸螺纹大小的比率与主轴转速和丝杠转速的比率相同。

Determining the most advantageous feeds and speeds for a particular lathe operation depends on the kind of material being worked on, the type of tool, the diameter and the length of the workpiece, the type of cut desired (rough or finishing), the cutting oil used, and the condition of the lathe being used.

确定针对某一特定机床操作的最佳的进给速度和切削速度取决于被加工的材料类型、道具类型、工件的直径和长度、期望的切削类型（粗加工还是精加工）、所使用的切削油，以及所使用车床的条件。

Any number of threads can be cut by merely changing the gear in the connecting gear train to obtain the desired ratio of the spindle and the lead screw speeds.

只要变换所连齿轮系中的齿轮获得预期的主轴和丝杠转速比，就可以切削任意螺距的螺纹。又如，rotating workpiece（旋转工件），clamping lever（锁紧手柄），work of varying length（变化长度的工件）等。

3. 不定式作定语

To determine the rotational speed necessary to produce a given cutting speed, it is necessary to know the diameter of the workpiece to be cut.

为了确定产生给定的切削速度所需的转速，有必要知道待切削工件的直径。

Since this form of CNC machine can perform multiple operation in a single program (as many CNC machines can), the beginner should also know the basics of how to process workpieces machined by turning so a sequence of machining operations can be developed for workpieces to be machined.

现在分词作定语后置多表示正在进行，过去分词作定语后置多表示已完成的动作或被动句意，而动词不定式作定语后置表示要做的事尚未发生，常与名词有动宾关系，形式为不及物动词加介词。以上3种定语形式经常翻译为"所……"。

4. 形容词做后置定语

To determine the rotational speed necessary to produce a given cutting speed, it is necessary to know the diameter of the workpiece to be cut.

为了确定给定的切削速度所需要的转速，有必要知道待切削工件的直径。rotational speed necessary 相当于 rotational speed which is necessary。

5. 定语从句

The spindle is the main rotating shaft on which the chunk is mounted, meaning that the workpiece revolves with it.

主轴是装有卡盘的主要旋转轴，即工件和主轴一起旋转。对于较短的定语从句翻译成"……的"。

A lathe is a machine that turns a piece of metal round and round against a sharp tool.

车床是一种用相对尖锐的刀具连续转动金属材料的机床。通常定义、解释某事物时用到定语从句。

Listening

Task I Listen to 2 conversations and choose the best answer.

Conversation 1

1. A. Breakfast.　　　　　　　　B. Dinner.
   C. A 5-dollar gift card.　　　　D. Bus service to the airport.
2. A. His member card.　　　　　B. His driving license.
   C. His credit card.　　　　　　D. His passport.

Conversation 2

3. A. The telephone is out of order.   B. The line is busy.
   C. He is at a meeting.   D. He won't be back until next Monday.
4. A. It has been cancelled.   B. It will arrive on time.
   C. It has been delayed.   D. It will arrive ahead of schedule.
5. A. Make an appointment with her.
   B. Talk with her about a new order.
   C. Send her an e-mail about the shipment.
   D. Call her back when receiving the shipment.

Task II Listen to the passage and fill in the blanks with the missing words or phrases.

Some managers have noticed recently that the employees in the company are taking advantage of the policy of having breaks. The workers have two 15-minute breaks per ___1___ . However, the two breaks are lasting ___2___ as 25 to 30 minutes each. The workers complain that the work is so ___3___ that they need longer breaks. Also the dining hall is so ___4___ that it takes too long to walk there and back. But the company is losing hundreds of work hours each year. Should employees be paid for the time they are not working? The general manager has to call a meeting to ___5___ this matter.

Extensive Reading

Directions: After reading the passage, you are required to complete the outline below briefly.

Human flesh search engine is a Chinese term for the research using Internet media such as blogs and forums. The system is based on massive human collaboration. It has generally been regarded as being for the purpose of exposing individual privacy to the public. More recent analyses, however, have shown that it is also used for a number of other reasons, including exposing government corruption, identifying hit-and-run drivers, and exposing scientific fraud, as well as for more "entertainment" related items such as identifying people seen in pictures.

Because of the convenient and efficient nature of internet, the human flesh search is often used to acquire information usually difficult or impossible to find by other conventional means (such as a library or web search engines). Such information, once available, can be rapidly distributed to hundreds of websites, making it an extremely powerful mass medium. Because personal knowledge or unofficial (sometimes illegal) access are frequently depended on to acquire this information, the reliability and accuracy of such searches often vary.

Human Flesh Search Engine

1. The human flesh search engine is based on ___1___ .
2. Its purpose is seen as exposing ___2___ to the public.
3. It's also used for exposing ___3___ , identifying ___4___ and exposing ___5___ .
4. It's used to acquire information that's ___6___ to find by conventional means.
5. Because we frequently depend on ___7___ or ___8___ to acquire the information, the reliability and accuracy of such searches often vary.

Section Five Writing

Speeches（致辞）

致辞是一种礼仪性应用文体，是指在一些社交场合中，如宴会、典礼、开幕式和闭幕式、学术谈论会等，向对方表示欢迎、祝贺、答谢等情感的简单讲话，起着增进友谊、加深了解和促进交流的作用。致辞的主题鲜明、目的明确、语言生动、层次分明、词句精练、全文紧凑、热情洋溢。

Sample

An Opening Speech

Message from Mayor of Dalian,
Honorary President Organizing Committee
of the 11th Dalian Fashion Festival

Ladies and Gentlemen,

Friends from all over the world,

It is a great honor for the city of Dalian to hold the 11th Dalian Fashion Festival. I, on behalf of the Dalian government and the Dalian people, extend my warm welcome to the friends and businessmen throughout the world, who have come to Dalian.

With the development of the cooperation and trade, the Dalian Fashion Festival has become an indispensable part of our social life. It has been known to all that through the festival, Dalian is full of vitality.

The friends from all parts of the world have come here to cooperate with each other, strengthening our mutual understanding, developing friendship, and promoting the development of international economy and trade. Both the government and the people of Dalian feel heartily happy to have the opportunity to make our contributions to promote the development of the economy and trade, friendship and progress among us.

I sincerely wish Dalian Fashion Festival complete successfully.

May this festival write a new page!

开幕词

第十一届大连国际服装节组委会
名誉主席、大连市市长的致辞

女士们，先生们，来自世界各地的朋友们：

　　大连荣幸地举办第十一届大连国际服装节，我谨代表大连政府和大连人民，对光临大连的各国朋友和商人表示热烈的欢迎。

　　随着合作和贸易的发展，大连国际服装节已经成为我们社会生活不可缺少的组成部分。众所周知，通过举办该服装节，大连充满了勃勃的生机。

　　来自世界各地的朋友欢聚在这里，相互合作，加强了相互了解，发展了友谊，促进了国际经济和贸易的发展。大连政府和大连人民为有此良机促进我们之间的经济和贸易、友谊和进步的发展做出我们的贡献而感到由衷的高兴。

　　我真诚地祝愿本届服装节圆满成功。

　　祝服装节谱写新的篇章！

致辞格式

　　常见的致辞包括欢迎词、欢送词、祝酒词、答谢词、告别词、开幕词、闭幕词等。致辞的种类不同，形式各异，一般包括以下几个部分。

　　1. 开篇称呼：致辞开篇一定要有称呼，如女士们、先生们、朋友们等。

　　2. 引言：引言部分是对到场的宾朋（或是对主人的欢迎或欢送）表示感谢。

　　3. 正文：正文部分根据特定情况，或介绍活动的原因，事情的安排，做好某事的愿望，或赞扬对方的才华、功绩，或强调双方的关系，或是表达感激之情等。

　　4. 结束语：结束语部分是再度表示欢迎、感谢或良好祝愿之类的言辞。

致辞常用句型

It is a great pleasure for us to have you all visit our company.
我们非常高兴诸位能来我公司访问。

I'd like to extend a warm welcome to you all.
我要向各位致以热烈的欢迎。

We are particularly grateful to all of you, who have travelled long distances in order to participate in this meeting.
我们要特别感谢远道而来参加会议的人们。

On behalf of our company, I would like to say how delighted we are to welcome you here.
作为公司代表，我想说能在这里欢迎你们我特别高兴。

I'm greatly honored to speak at this welcoming party.
我为能在欢迎会上讲话而深感荣幸。

I hope your trip to Beijing will mark the beginning of a long-standing cooperation between us.
我希望你们这次北京之行将标志着我们之间长期合作的开始。

Please come to visit our company if time permits.
如果时间允许的话，请来参观我们的公司。

We very much appreciate your efforts in making our visit successful.
我们非常感谢各位为我们访问成功所付出的努力。
Please join me in drinking a toast to the health and happiness of one and all throughout the year.
让我们举杯祝福新的一年大家身体健康,心情愉快!
No words can describe our friendship and joy we felt on this visit.
言语无法表达我们的友谊,无法描述我们这次访问之愉快。
Please take good care of yourselves, and I wish you continued success.
请各位多多保重,祝各位顺遂。
I'm pleased to announce the recipients of this year's Award for Outstanding Performance.
我很高兴地宣布今年杰出表现奖的获奖者。

Practice

Please write a speech according to the following information in Chinese.

新年致辞

女士们,先生们:

新年快乐!

这是我们公司成立之后的第一个新年,回想这一年,大家的努力使我们超额完成了年度计划,再次感谢各位。

我希望今年公司和各位同仁继续成长。最后我要祝大家身体健康,幸福快乐!

Section Six Grammar

Adverbial Clause I 状语从句(一)

状语从句(Adverbial Clause)状语从句指句子用作状语时,起副词作用的句子。它可以修饰谓语、非谓语动词、定语、状语或整个句子。根据其作用可分为时间、地点、原因、条件、目的、结果、让步、方式和比较状语从句等。状语从句一般由连词(从属连词)引导,也可以由词组引起。从句位于句首或句中时通常用逗号与主句隔开。

一、时间状语从句

常用引导词:when, as, while, as soon as, before, after, since, till, until

特殊引导词:the minute, the moment, the second, every time, the day, the instant, immediately, directly, no sooner... than, hardly... when, scarcely... when 在时间状语从句中,要注意时态一致。一般情况下主句是将来时的时候,从句要用一般现在时。

Mozart started writing music when he was four years old.
(当)莫扎特4岁的时候,开始作曲。
He left the classroom after he had finished his homework the other day.
他前几天做完作业之后才离开教室。
I will write to you as soon as I get home.
我一到家就给你写信。

Mr Green has taught in that school since he came to China three years ago.
格林先生3年前来中国开始，就在这所学校教书。

注意：

1. when, while 和 as 的区别

when 引导的从句的谓语动词可以是延续性的动词，又可以是瞬时动词。并且 when 有时表示"就在那时"。例如：

When she came in, I stopped eating. 她进来时，我停止吃饭。（瞬时动词）

When I lived in the countryside, I used to carry some water for him. 当我住在农村时，我常常为他担水。（延续性的动词）

We were about to leave when he came in.
我们正要离开时他进来了。

While 引导的从句的谓语动词必须是延续性的，并强调主句和从句的动作同时发生（或者相对应）。while 有时还可以表示对比。例如：

While my wife was reading the newspaper, I was watching TV.（was reading 是延续性的动词，was reading 和 was watching 同时发生）

I like playing football while you like playing basketball. 我喜欢踢足球，而你喜欢打篮球。（对比）

As 表示"一边……一边……"，as 引导的动作是延续性的动作，一般用于主句和从句动作同时发生；as 也可以强调"一先一后"。例如：

We always sing as we walk.
我们总是边走边唱。（as 表示"一边……一边……"）

As we was going out, it began to snow.
当我们出门时，开始下雪了。（as 强调句中两个动作紧接着先后发生，而不强调开始下雪的特定时间）

2. 由 till 或 until 引导的时间状语从句

till 和 until 一般情况下两者可以互换，但是在强调句型中多用 until。并且要注意的是：如果主句中的谓语动词是瞬时动词，必须用否定形式；如果主句中的谓语动词是延续性动词，用肯定或否定形式都可以，但表达的意思不同。例如：

I didn't go to bed until（till）my father came back.
直到我父亲回来我才上床睡觉。

It was not until the meeting was over that he began to teach me English.
直到散会之后他才开始教我英语。

I worked until he came back. 我工作到他回来为止。

I didn't work until he came back. 他回来我才开始工作。

Please wait until I arrived. 在我到达之前请等我。

3. 由 since 引导的时间状语从句

since 引导的从句的谓语动词可以是延续性的动词，也可以是瞬时动词。一般情况下，从句谓语动词用一般过去时，而主句的谓语动词用现在完成时。但在"It is + 时间 + since"从句的句型中，主句多用一般现在时。例如：

I have been in Beijing since you left. 自从你离开以来，我一直在北京。

It is four years since my sister lived in Beijing.

我妹妹住在北京已经 4 年了。

It is five months since our boss was in Beijing.

我们老板离开北京有 5 个月了。

4. hardly(scarcely, rarely)…when / before/no sooner…than

主句用过去完成时，从句用一般过去时。当 hardly, scarcely, rarely 和 no sooner 位于句首时，主句应用倒装语序。例如：

He had no sooner arrived home than he was asked to start on another journey. 他刚到家，就被邀请开始另一旅程。

Hardly had I sat down when he stepped in. 我刚坐下，他就进来了。

He had hardly fallen asleep when he felt a soft touch on his shoulder.

他刚要入睡就感到肩膀上被轻轻一触。

5. 由 by the time 引导的时间状语从句

注意时态的变化：在一般情况下，如果从句的谓语动词用一般过去时，主句的谓语动词用过去完成时；如果主句的谓语动词用一般现在时，主句的谓语动词用将来完成时。例如：

By the time you came back, I had finished this book.

到你回来时，我已经写完这本书了。

By the time you come here tomorrow, I will have finished this work.

你明天来这儿的时候，我将已经完成此工作了。

二、地点状语从句

常用引导词：where

特殊引导词：wherever, anywhere, everywhere

Where there is no rain, farming is difficult or impossible.

在没有雨水的地方，耕作是困难的，或根本不可能的。

You should have put the book where you found it.

你本来应该把书放回原来的地方。

Wherever you go, I go too.

无论你到什么地方，我都去。(wherever=no matter where)

三、条件状语从句

常用引导词：if, unless

特殊引导词：as/so long as, only if, providing/provided that, supposing that, in case that, on condition that

If it doesn't rain tomorrow, we will go hiking.

如果明天不下雨，我们就去远足。

I will go to the party unless he goes there too.

我不会去参加聚会的，除非他也去。（如果他不去，我也不去。）

You will be late unless you leave immediately.

如果你不马上走，你将会迟到的。(=If you don't leave immediately, you will be late.)

注意：用条件状语从句时要注意时态的正确使用，当主句是将来时的时候，从句要用一般现在时。

He will not leave if it isn't fine tomorrow.
如果明天天气不好，他就不走了。

四、原因状语从句

常用引导词：because, since, as, for

特殊引导词：seeing that, now that, in that, considering that, given that

I didn't go to school yesterday because I was ill.

我昨天没去上学，因为我生病了。

Since everybody is here, let's begin our meeting.

既然大家都来了，让我们开始开会吧。

because, since, as 辨析：because 语势最强，用来说明人所不知的原因，回答 why 提出的问题。当原因是显而易见的或已为人们所知，就用 as 或 since。

I didn't go, because I was afraid. 我不去，因为我害怕。

Since /As the weather is so bad, we have to delay our journey.

由于天气太糟糕了，我们不得不推迟行程。

Exercises

Choose the best answer.

1. I'll let you know _____ he comes back.
   A. before   B. because   C. as soon as   D. although
2. She will sing a song _____ she is asked.
   A. if   B. unless   C. for   D. since
3. It is about ten years _____ I met you last.
   A. since   B. for   C. when   D. as
4. I learned a little Russian _____ I was at middle school.
   A. though   B. although   C. as if   D. when
5. _____ we got to the station, the train had left already.
   A. If   B. Unless   C. Since   D. When
6. _____ the rain stops, we'll set off for the station.
   A. Before   B. Unless   C. As soon as   D. Though
7. We didn't go home _____ we finished the work.
   A. since   B. until   C. because   D. though
8. I'll stay here _____ everyone else comes back.
   A. even if   B. as though   C. because   D. until
9. You'll miss the train _____ you hurry up.
   A. unless   B. as   C. if   D. until
10. When you read the book, you'd better make a mark _____ you have any questions.
    A. at which   B. at where   C. the place   D. where
11. We'd better hurry _____ it is getting dark.
    A. and   B. but   C. as   D. unless

12. I didn't manage to do it _____ you had explained how.
    A. until   B. unless   C. when   D. before
13. _____ he comes, we won't be able to go.
    A. Without   B. Unless   C. Except   D. Even
14. You may arrive in Beijing early _____ you mind taking the night train.
    A. that   B. though   C. unless   D. if
15. _____ we stood at the top of the building, the people below were hardly visible.
    A. As   B. Although   C. Unless   D. In spite of
16. Scarcely was George Washington in his teens _____ his father died.
    A. than   B. as   C. while   D. when
17. The house stood _____ there had been a rock.
    A. which   B. at which   C. when   D. where
18. _____ the day went on, the weather got worse.
    A. With   B. Since   C. While   D. As
19. —What was the party like?
    —Wonderful. It's years _____ I enjoyed myself so much.
    A. after   B. when   C. before   D. since
20. After the war, a new school building was put up _____ there had once been a theatre.
    A. that   B. where   C. which   D. when

CULTURE TIPS

扫一扫了解常用搜索引擎

Unit Nine

Network and System Administrator

Section One Warming Up

Pair work

System and network administrators play an important role in maintaining the regular operation of a company. Understand some of their responsibilities listed below and match the pictures with the statements.

A. Set up security policies for users
B. Perform backups of data.
C. Quickly arrange repair for hardware in occasion of hardware failure
D. Keep the network up and running
E. Apply operating system updates
F. Install and configure new hardware/software
G. Answer technical queries
H. Document the configuration of the system
I. Troubleshoot any reported problems
J. Manage password and identity
K. Tune system performance

1._____

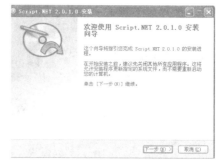
2._____

3._____

4._____

Section Two Real World

Listen and act the following dialogue which is about assigning network address.

Helping assign network address

Rex: Nice to meet you, Candy. I'm required to set up the computer system and the network for you.

Candy: Nice to meet you, too, Rex. Thank you for your help.

(After 20 minutes)

Rex: I've **customized** Windows to meet your needs. And this is your user **password** of this computer.

Candy: Thanks a lot. Can I access the Internet now?

Rex: Wait a moment. I will assign a **static** IP address for you.

Candy: The manager tells me this office is within the **coverage** of a **wireless** network. Isn't it more **flexible** and convenient?

Rex: Yes, but its speed is much slower than **wired** connection. **Moreover**, it can be easily affected by the **surroundings** and has some security risks.

Candy: Em, I see. Would you please **install** the PDF reader for me?

Rex: I've already installed it for you. You can find the **shortcuts** on your computer **desktop**. You can access the network now. If you meet any other problem, feel free to contact me.

Candy: Thank you very much.

Rex: My pleasure.

Words & Expressions

assign [ə'saɪn]	vt.	give sth to sb as a share of work to be done or of things to be used 分配；交给
customize ['kʌstəmaɪz]	vt.	make according to requirements 定制
password ['pɑːswɜːd]	n.	a secret word or phrase known only to a restricted group 密码
static ['stætɪk]	adj.	not moving, changing, or developing 静态的
coverage ['kʌvərɪdʒ]	n.	extent to which sth is covered 覆盖范围
wireless ['waɪələs]	adj.	having no wires 无线的
flexible ['fleksəbl]	adj.	easily changed to suit new conditions 灵活的；可变通的
wired ['waɪəd]	adj.	equipped with wire or wires especially for electric or telephone service 有线的
moreover [mɔːr'əʊvə(r)]	adv.	besides what has been said 再者；此外；而且
surroundings [sə'raʊndɪŋz]	n.	the objects around; environment （周围的）环境（事物）
install [ɪn'stɔːl]	vt.	set up, ready for use 安装
shortcut ['ʃɔːtkʌt]	n.	a route shorter than the usual one 捷径；快捷键
desktop ['desktɒp]	n.	(computer science) the area of the screen in graphical user interfaces against which icons and windows appear 桌面
set up		建立；设立
meet one's needs		符合某人的需要
IP address		IP 地址（Internet Protocol Address 网际协议地址）
feel free		请便；不用客气

Find Information

Task I Read the dialogue carefully and then answer the following questions.

1. What's Rex required to do?

___

2. Is Candy's office covered with wireless network?

___

3. What's the disadvantage of wireless network compared with wired connection?
_____

4. What software does Rex install for Candy?
_____

5. Do you think Candy is pleased with Rex's service?
_____

Task II Read the dialogue carefully and then decide whether the following statements are true (T) or false (F).

(　) 1. Candy is coming to assign the network address for Rex.
(　) 2. Candy can access the Internet because there is wireless network in the office.
(　) 3. Wireless network is less convenient and flexible than wired connection.
(　) 4. When you surf the Internet through wireless network, there're some security risks.
(　) 5. Candy can find the shortcuts of PDF reader on her computer desktop.

Words Building

Task I Choose the best answer from the four choices A, B, C and D.

(　) 1. This software is _____ more useful than that one.
　　A. very　　　　　　B. much　　　　　　C. a lot of　　　　　D. less
(　) 2. He _____ Beijing for many times.
　　A. has been to　　　B. has gone to　　　C. have been to　　　D. have gone to
(　) 3. –Would you please not smoke here? Look at the sign. – _____ .
　　A. No, I will　　　　B. Yes, I will　　　　C. Sorry, I will　　　D. Sorry, I won't
(　) 4. I must _____ my lawyer before I make my final decisions.
　　A. contract　　　　B. contrast　　　　　C. contact　　　　　D. context
(　) 5. –Aren't you from Dandong?
　　– _____ . I'm from Shenyang.
　　A. Yes, I'm not.　　B. Yes, I am.　　　　C. No, I'm not.　　　D. No, I am.

Task II Fill in each blank with the proper form of the word given.

1. The hot _____ is caused by overload. (wireless)
2. The house was _____ by high walls. (surroundings)
3. You can deal with it _____ . (flexible)
4. A manager should be _____ to his staff. (access)
5. The workers are _____ a heating system. (install)

Task III Match the following English terms with the equivalent Chinese.

A — account lockout
B — denial of service
C — LAN
D — mutlidrop connection
E — IP address
F — terminal server
G — WAN
H — host
I — network troubleshooting
J — domain name

1. (　) 拒绝服务
2. (　) IP 地址
3. (　) 主机
4. (　) 多点连接
5. (　) 局域网
6. (　) 域名
7. (　) 网络故障诊断与维修
8. (　) 账号封锁
9. (　) 广域网
10. (　) 终端服务器

Cheer up Your Ears

Task I Listen and write down what you've heard. Then read and recite till you can use them fluently.

1. I'm required to _____ the computer system and the network for you.
2. I've customized Windows to _____ your needs.
3. A network is a group of two or more _____ linked together.
4. This is your user _____ of this computer.
5. Security management protects a network from unauthorized _____ .
6. Wireless network can be easily affected by the _____ and has some security risks.
7. Would you please _____ the PDF reader for me?
8. You can find the shortcuts on your computer _____ .
9. The ability to _____ network devices is fundamental to their utility.
10. If you meet any other problem, _____ to contact me.

Task II Listen and fill in the blanks with the missing words and role play the conversation with your partner.

A: I've got a problem.
B: Don't worry. What's the matter?
A: My printer can't __1__ . And I still have a pile of documents to print.
B: Let me have a look. (Reading words on the screen) "Printer selected is not valid". Is this the default printer?
A: Yes. I __2__ all the instructions when installing the printer.
B: The cable is properly connected. Did you reset it by __3__ ?
A: I've tried three times, but the problem is still there.
B: The printer driver needs to be reinstalled. It will take a while.

A: But I can't wait. I must hand in the contract ___4___ .
B: Don't worry. I'll ___5___ your computer to a printer on the network. Then you can use the printer in the Manager's Office.
A: Thank you so much.
B: You're welcome.

Task III Listen to 5 recorded questions and choose the best answer.

1. A. Hurry up.   B. Nothing left.
   C. No more.   D. Me, too.
2. A. It cost me 20 dollars.   B. It was a waste of time.
   C. It took me 3 hours.   D. It's time for dinner now.
3. A. 14 dollars.   B. It's next to the station.
   C. Sorry, I have no money.   D. Sorry, I don't know the way.
4. A. You're right.   B. But it's still early.
   C. What do you think of it?   D. Why are you so late?
5. A. Don't you know?   B. Thanks a lot.
   C. No problem.   D. It doesn't matter.

Table Talk

Task I Complete the following dialogue by translating Chinese into English orally.

A: (on the phone) Is that the Front Office?
B: Yes. What can I do for you, sir?
A: I can't surf online with my laptop computer. Could you help me?
A: What is your room number, sir? Our network administrator will ___1___ （查看）it right now.
  (A few minutes later)
A: (Knocking at the door) Excuse me, sir. I'm the network administrator of the hotel. May I come in?
C: Yes, please. I don't know ___2___ （怎么了）with the network. My computer works well in other hotels.
A: May I check the ___3___ （网络连接）of your computer?
C: Of course.
A: Oh, I see. Your laptop is configured with a static IP address. Would you like to ___4___ （写下它）? I will set your computer to pick up an IP address automatically from a predefined range of address.
C: (Writing down the IP address) OK, here you are.
A: (Modifying the setting) You can access the network now. But since you are visiting a foreign server, the speed is ___5___ （有点）slow.

C: That's all right. Thanks a lot.
A: My pleasure. Wish you a pleasant stay in our hotel.

Task II **Pair-work. Role-play a conversation about network management with your partner.**

Situation

Imagine you can't surf online. You want to ask your friend for some advice. Now you are talking with your friend.

Section Three Brighten Your Eyes

Pre-reading questions:
1. Who's responsible for maintaining the network?
2. Do you know what an administrator's job like?

Network and System Administrators

Network and system administrators can be found in many **environments**, including industry, government, **manufacturing**, and education. They normally work in comfortable offices or computer **laboratories**.

A typical work week is forty hours; however many administrators tend to be "on call" for extended evening or weekend duty. Overtime is not uncommon due to the possibility of unexpected computer or network **technical** problems **arising**.

SYSTEM ADMINISTRATOR

Administrators who work as **consultants** can be away from their offices and may need to work in a **client's** office for weeks or even months at a time. Administrators are now often able to provide technical **support** from **remote** locations, **reducing** the need to travel to a customer's workplace.

The job tends to be interesting and **generally** well-paying. For those who type on a keyboard for long periods **drawbacks** include back **discomfort**, **eyestrain**, and hand/wrist problems. Frequent **deadlines** and **accuracy** and **precision** can be causes for **stress**. Also, the need to **constantly interact** with people, while satisfying, also can be frustrating at times, particularly if the computer systems are not working like what they are supposed to.

Words & Expressions

administrator [əd'mɪnɪstreɪtə(r)]   n.   someone who administrate a business or

			manages a government agency or department 管理人；行政官
environment	[ɪnˈvaɪrənmənt]	n.	conditions, circumstances, etc. affecting people's lives 环境；周围状况
manufacture	[ˌmænjuˈfæktʃə(r)]	vt.	make, produce (goods, etc.) on a large scale by machinery 制造
laboratory	[ləˈbɒrətri]	n.	a special building or room in which a scientist works to examine, test, or prepare materials 实验室
technical	[ˈteknɪkl]	adj.	having or giving special and usually practical knowledge 技术的；工艺的
arise	[əˈraɪz]	vt.	appear 出现，发生
consultant	[kənˈsʌltənt]	n.	a person who gives professional advice to others 顾问
client	[ˈklaɪənt]	n.	a person who pays another person, for help or advice 委托人；customer 顾客
support	[səˈpɔːt]	n.	supporting or being supported 支持；赞助
		vt.	keep from falling; hold up; help by agreeing with 支撑；支持
remote	[rɪˈməʊt]	adj.	far away from other communities, house, etc.; isolated 遥远的；远程的
reduce	[rɪˈdjuːs]	v.	make less in size, amount, price, degree, etc. 缩减；减少；降低
generally	[ˈdʒenərəli]	adv.	usually；in general 通常；大体上
drawback	[ˈdrɔːbæk]	n.	disadvantage, problem 缺点；不利条件
discomfort	[dɪsˈkʌmfət]	n.	an uncomfortable feeling of mental painfulness or distress 不适
eyestrain	[ˈaɪstreɪn]	n.	a tiredness of the eyes caused by prolonged close work of a person with an uncorrected vision problem 眼睛疲劳
deadline	[ˈdedlaɪn]	n.	a date or time before which sth. must be done or completed 最后期限
accuracy	[ˈækjərəsi]	n.	the state of being exact or correct; the ability to do sth skillfully without making mistakes 准确（性）；精确（度）
precision	[prɪˈsɪʒn]	n.	exactness 精确，准确
stress	[stres]	n.	pressure caused by the problems of living, too much work, etc. 压力；紧张
constantly	[ˈkɒnstəntli]	adv.	without interruption 不断地
interact	[ˌɪntərˈækt]	v.	have an effect on each other or sth else by

	being or working closely together 相互作用（影响）；互相配合
tend to	易于做某事；有做……的倾向
on call	随时可用，随时到
due to	因为；由于
at a time	每次；一次
at times	有时；偶尔
be supposed to	应该

Find Information

Task I Read the passage carefully and then answer the following questions.

1. What kind of environment do administrators work in?

2. How long do administrators work typically a week?

3. Is overtime work common for administrators? Why?

4. Why is there no need for administrators to travel to a customer's workplace?

5. What may be the health problem of administrators?

Task II Read the passage carefully and then decide whether the following statements are true (T) or false (F).

( ) 1. Network and system administrators need to work overtime because of lack of efficiency.
( ) 2. Administrators must work in their own offices all the time.
( ) 3. The salary of administrators is not bad.
( ) 4. Administrators may have problems with their back.
( ) 5. Administrators need to interact with people constantly.

Words Building

Task I Translate the following phrases into English.

1. 网络和系统管理员_____
2. 机房_____
3. 随叫随到_____
4. 技术问题_____

5. 技术支持_____
6. 远程_____
7. 长时间打字_____
8. 背部不适_____

Task II **Translate the following sentences into Chinese or English.**

1. Network and system administrators can be found in many environments.
_____

2. Administrators are now often able to provide technical support from remote locations.
_____

3. Administrators need to make sure that all components（组件）of a network are working together properly.
_____

4. We should always consider the operation of the whole network.
_____

5. Network and computer system administrators are professionals who are responsible for designing, installing, and supporting computer systems within an organization.
_____

6. 担任客户咨询顾问工作的管理人员可以不待在自己的办公室。
_____

7. 这项工作往往很有趣并且薪水一般来说也很优厚。
_____

8. 由于网络技术问题，管理人员必须加班。
_____

9. 管理人员不必前往顾客的工作地点提供技术支持。
_____

10. 网络和系统管理人员需要不断地与他人进行交流。
_____

Task III **Choose the best answer from the four choices A, B, C and D.**

(　　) 1. —I don't know if he _____ .
　　　　—He will come if it _____ .
　　　A. comes, won't rain　　　　　　　B. will come, doesn't rain
　　　C. comes, doesn't rain　　　　　　D. will come, won't rain

(　　) 2. The failure of the scheme was _____ bad management.
　　　A. instead of　　B. due to　　C. thanks for　　D. such as

(　　) 3. We will go to watch the movie about _____ you talked yesterday.
　　　A. that　　B. where　　C. who　　D. which

(　　) 4. This sentence needs _____ .

A. improves   B. improved   C. improving   D. improve
( ) 5. He was supposed _____ us advice, but all he came up with were airy-fairy ideas.
A. give   B. giving   C. to give   D. gives

Task IV Fill in each blank with the proper form of the word given.

1. She usually writes about _____ issues. (environment)
2. I need to _____ with my colleagues on the proposals. (consultant)
3. When you're not _____ , you can be more productive. (stress)
4. The dress is carefully styled for maximum _____ . (discomfort)
5. The _____ friction between the young couple finally caused divorce. (constantly)

Cheer up Your Ears

Task I Listen and write down what you've heard. Then read and recite till you can use them fluently.

1. Is there any _____ that makes you feel it difficult to manage your work?
2. Technical knowledge is a _____ , but so are good social and communication skills.
3. Network and system administrators can be found in many _____ .
4. I'm required to help as many users as _____ .
5. Administrators who work as _____ can be away from their offices.
6. Administrators need to constantly _____ with people.
7. The specific _____ of a system administrator tend to vary widely from one organization to another.
8. Many administrators tend to be "on call" for _____ evening or weekend duty.
9. Network _____ is a latent issue that computer and network users tend to ignore.
10. Most network _____ positions require a breadth of technical knowledge.

Task II Listen to 5 short dialogues and choose the best answer.

1. A. The travel agency is at the airport.
   B. The theater is near Broadway.
   C. The woman is going on a bus trip.
   D. The woman bought an airplane ticket.
2. A. Traveler and travel agent.
   B. Host and guest.
   C. Fright attendant and passenger.
   D. Lecturer and student.
3. A. They'll have to go to a later show.
   B. The people in line all have tickets.
   C. She doesn't want to go to the second show.

D. They won't have to wait much longer.
4. A. Booking an airline status.　　B. Their social status.
   C. Their business.　　D. The man's preference.
5. A. Airline tickets.　　B. Concert tickets.
   C. Movie tickets.　　D. Train tickets.

Table Talk

Group Work

Talk about the job of network administrators with your partner as much as you know.

_____

_____

Section Four Translation Skills

科技英语的翻译方法与技巧——被动语态的翻译

科技英语中大量使用被动语态，这是因为文章需要客观地叙述事理，而不是强调动作的主体。

The clutch controls the direction of spindle rotation.
离合器控制主轴旋转方向。
The direction of spindle rotation is controlled by the clutch.
主轴旋转方向由离合器控制。

在第一句中，动词的主语 the clutch 实施动词表示的动作，这个动词处于"主动语态"；在第二句中，动词的动作被施加到主语上，该动词处于"被动语态"。

只有及物动词（其后可跟宾语）可用于被动语态。主动形式的宾语，通常是被动形式的主语。

1. 被动语态的种类
各种时态的被动语态如下：
（1）一般现在时的被动语态：
Computer is used in business, governments and institutions.
计算机用于商业、政府和公共机构。
（2）现在进行时的被动语态：
Computers are being used in machine tool control.
计算机正用于机床的控制。
（3）过去进行时的被动语态：
The hard-wired NC was being used widely in 1960s.
20世纪60年代时，硬线数控应用很广。
（4）不定式的被动语态：

If a large amount of stock is to be removed, it is advisable to take one or more roughing cuts and then take light finishing cuts at relatively light speed.

如果毛坯切削量很大，建议进行一次以上的粗切削，然后以相对高速进行少量精切削。

（5）带情态动词的被动语态：

Finishing cuts are generally very light, therefore, the cutting speed can be increased since the chip is thin.

精切削一般量很小，由于切屑很薄，因此可以提高切削速度。

（6）短语动词的被动语态：

The documents attached to the machine should be taken good care of.

必须妥善保管这些随机资料。

2. 被动语态的翻译方法

把英语被动句译成汉语时，一般可以采用下列处理方法。

（1）译成汉语被动句

用"被""由""受""靠""给""遭"等汉语中表达被动概念的介词引导出施动者。

The power feed is engaged by the longitudinal/cross feed level on the apron.

自动进给由溜板箱上的纵向/横向进给手柄啮合。

The diameter of the workpiece is determined by a caliper or micrometer.

工件直径由卡尺或千分尺确定。

（2）译成汉语无主语句

Good safety practices should be followed to ensure safe machining.

必须遵守良好的安全规程以保证安全加工。

（3）增译"人们""有人""操作员"等主语

It is recommended that you never remove your hand from the chunk key when it is in the chunk.

我们建议只要卡盘扳手在卡盘上，手就不要离开扳手。

（4）由 by 或 in 引导的状语往往可以转换为汉语的主语

M codes are commonly used by the machine tool builder to give the user programmable ON/OFF switches for machine functions.

机床制造商通常用 M 代码给用户提供可编程的机床开/关功能。

Information regarding the machine's construction is usually published right in the machine tool builder's manual.

机床制造厂的说明书通常发布机床结构方面的信息。

（5）将英语句子中的一个适当成分译成汉语中的主语

有时英文里的被动含义在中英文里不一定需要表示出来。

The cutting tool is moved a definite distance along the work for each revolution of the spindle.

主轴每转一转，刀具沿工件移动固定的距离。

The apron is attached to the front of the carriage.

溜板箱连接在大拖板的前方。

3. 常用句型

科技英语中有许多惯用句型，其中很多运用被动语态，常见的有：

The case is that...	问题在于……
It can be seen that...	可以看出……
It has been proved that...	已经证明……
It appears that...	看来……
Results demonstrate that...	结果证明……
It is necessary to point out that...	有必要指出……
It is not hard to imagine that...	不难想象……
It will be found that...	将会发现……
If follows that...	由此可见……
This implies...	意味着……
Of importance is (are)...	重要的是……
Of recent concern is (are)...	近来引起重视的是……
Particularly noteworthy is that...	特别值得注意的是……
It must be noted that...	必须指出……
It was reported that...	据报道……
It should be mentioned that...	应该指出……
We believe that...	我们认为……
Practice has shown that...	实践证明……
The fact is that...	事实是……
The purpose of this paper is...	本文的目的是……
Calculations indicated that...	预测表明……
Experience has shown that...	经验已证明……
Fig. 2 illustrates...	图2表明了……
I suppose that...	我以为……
It is evident that...	显然……
Some believe that...	有些人认为……
One can only say that...	只能认为……
Tests have proven that...	实验证明……
It is well known that...	众所周知……
It is possible that...	可能……
It may be remarked that...	可以认为……

Listening

Task 1 Listen to 2 conversations and choose the best answer.

Conversation 1

1. A. Stay in the city. B. Work with her father.
   C. Read books. D. Go to the beach with her family.
2. A. Go to New York. B. Go camping.

C. Work at the store.   D. Learn to work.

Conversation 2

3. A. In a reporter's office.   B. In a secretary's office.
   C. In a policeman's office.   D. In a gatekeeper's office.
4. A. A commercial Party.
   B. A press conference.
   C. An interview with the manager.
   D. An appointment with the secretary.
5. A. 13091462587.   B. 13901246578.
   C. 13910426875.   D. 13901426587.

Task II Listen to the passage and fill in the blanks with the missing words or phrases.

Tourism was not always as important as it is today. In the past, only ___1___ people could travel on vacation to other countries. But in ___2___ one person in ten visited a country away from home.

More people travel today because there is a ___3___ middle class in many parts of the world. People now have more money for travel. Special airplane fares for tourists make travel ___4___ and thus more attractive than ever before. One person does not travel for the same reason as another. But most people ___5___ seeing countries that are different from their own. They also like to meet new people and try new foods.

Extensive Reading

Directions: After reading the passage, you are required to complete the outline below briefly.

The primary function of a network and system administrator is to install, support, and maintain the components of a computer system, including servers and networking elements. In addition, administrators are typically responsible for maintaining cognizance of the newest technologies in order to determine how these can be configured to best serve the short and long term goals of their organizations. Responsibilities may include any or all of the following:

Maintenance of network hardware and software

Determination of system hardware and software requirements

Ensuring continuous network availability to all system users

Planning and implementation of network security measures

Recommendation of changes to improve systems and network configurations

Maintenance of parts inventory for repairs

Implementation of updates, patches, and configuration changes

Performance of routine backups

Coordination of computer network access and use

Responsibilities of Network and System Administrators

1. ___1___ of network hardware and software
2. Determination of system hardware and software ___2___
3. Ensuring continuous network ___3___ to all system users
4. Planning and ___4___ of network security measures
5. ___5___ of changes to improve systems and network configurations
6. Maintenance of parts inventory for ___6___
7. ___7___ of updates, patches, and configuration changes
8. ___8___ of routine backups

Section Five Writing

Certificate 证明

证明书是用来证明一个人的身份、学历、经历、婚姻状况、身体情况、受到奖励或聘请等与个人有关的事情真相的文件。常见的证明书包括身份证、毕业证、奖状、聘书、病情证明、公证书等。证明书的语言应正式、简洁。

Sample

Read and understand the following certificates.

Sample 1 Graduation Certificate

Graduation Certificate

This is to certify that Li Jing, female, aged 23, a student from Dandong, Liaoning Province, majoring in Chemistry in the Department of Applied Chemistry from September 2011 to July 2015, completed the four years' courses required by the syllabus, and graduated from Liaoning University.

President: (signature)

Liaoning University (seal)

Date: July 2, 2015

范例 1 毕业证书

毕业证书

兹证明李静（女，23岁）系辽宁省丹东市人，自2011年9月至2015年7月在我校应用化学系攻读化学专业，完成四年制学习，达到教学大纲要求，准予毕业。

校长：（签名）

辽宁大学（盖章）

2015年7月2日

Sample 2 Notarized Certificate

> Notarial Certificate of Birth
> （2005）No.718
>
> This is to certify that Wang Fei, male, was born on May 23, 1981 in Dandong City, Liaoning Province. His father's name is Wang Tieniu and his mother's name is Wu Lan according to the record of file kept by the Administration of Census Register of Dandong City, Liaoning Province.
>
> Notary: Liu Tong（signature）
> Dandong Notary Public Office Liaoning Province
> The People's Republic of China
> March 12, 2006

范例 2 公证书

> 出生公证书
> （2005）证字第 718 号
>
> 根据辽宁省丹东市户籍管理机关档案记载，兹证明王飞，男，于 1981 年 5 月 23 日在辽宁省丹东市出生。王飞的父亲是王铁牛，王飞的母亲是吴兰。
>
> 中华人民共和国辽宁省丹东市公证处
> 公证员：刘通（签名）
> 2006 年 3 月 12 日

证明书基本格式

证明信的写法通常也采用一般信件格式，但多省掉收信人的姓名、地址和结束用语，通常在信纸的上方先注明"Certificate"的字样。要证明的内容开门见山地写在下方，然后再写上署名和时间。署名和时间也可以安放在其他相应的位置。称呼多用"To Whom It May Concern"意即"有关负责人，敬启者"，但此项也常省略。写证明书要求言简意赅、用词严谨准确。

毕业证书、学位证书和结业证书应包括持证人的姓名、年龄、性别、在校时间和成绩是否合格等内容。

聘书应包括被聘人的姓名、应聘职务、聘期和颁发聘书的单位。

奖状应包括被奖励人的姓名、工作单位和所获奖项。

公证书应包括证书号、证明内容、公证机关名称、公证员姓名（签字）、公证时间。

证明书常用句型

This is to certify that Ding Li was duly admitted to the degree of Bachelor of Arts in the Pujiang University on July 16, 2004.

兹证明丁利于 2004 年 7 月 16 日正式取得浦江大学文学学士学位。

This is to certify Mr. Milton, male, aged 43, is suffering from acute appendicitis.

现证明密尔顿先生，男，43 岁，患有急性阑尾炎。

Li Lei and Ma Li are hereby married of their own free will.
李雷和马丽自愿结婚。
This is to certify that Tony Smith died of illness on April 17, 2006.
兹证明托尼．史密斯先生于 2006 年 4 月 17 日病故。
Completed all the courses as required and satisfied all the requirements.
主修课程合格
Zhang Zhen is given this Certificate of Graduation.
准予张震毕业。

Practice

Directions: You are required to complete the following certification by translating the Chinese given below. You should write the answer in the composition sheet.

Diploma

Xu Tingting, _____ （女性）, _____ （出生于）on September 12, 1976, _____ （籍贯在）of Shanghai, was an undergraduate student _____ （学习）Foreign Trade of Pujiang University from September, 1999 to July, 2002. She _____ （完成了） all the prescribed four-year-undergraduate courses, has passed all the examinations and is _____ （按期） to be a graduate of Pujiang University.

Signature _____
No. 1234567890
President of Pujiang University
Issued on July 22, 2002

Section Six Grammar

Adverbial Clause II 状语从句（二）

一、目的状语从句

常用引导词：so that, in order that

特殊引导词：lest, in case, for fear that, in the hope that, for the purpose that, to the end that

I'll run slowly so that you can catch up with me.（目的）

我将慢慢跑以至你能赶上我。

We shall let you know the details soon in order that you can/may make your arrangements.

不久我们将会让你知道详情，以便你们能够做出安排。（目的）

二、结果状语从句

常用引导词：so... that, such... that,

特殊引导词：to the degree that, to the extent that, to such a degree that

He always studied so hard that he made great progress.

他总是那么努力，结果他取得了很大的进步。

It's such nice weather that all of us want to go to the park.

天气是如此的好，我们大家都想去公园玩。

比较：so 和 such

such 是形容词，修饰名词或名词词组，so 是副词，只能修饰形容词或副词。so 还可与表示数量的形容词 many, few, much, little（这 4 个形容词表多或表少时）连用，形成固定搭配。

so foolish　　　　　　　　such a fool
so nice a flower　　　　　　such a nice flower
so many / few flowers　　　such nice flowers
so much / little money.　　 such rapid progress
so many people　　　　　　such a lot of people

（so many 已成固定搭配，a lot of 虽相当于 many，但 a lot of 为名词性的，只能用 such 搭配。）

so...that 与 such...that 之间的转换既为 so 与 such 之间的转换。

The boy is so young that he can't go to school.
He is such a young boy that he can't go to school.
这个男孩太小还不能去上学。

三、让步状语从句

常用引导词：though, although, even if, even though

特殊引导词：as（用在让步状语从句中必须要倒装），while（一般用在句首），no matter...，in spite of the fact that, while, whatever, whoever, wherever, whenever, however, whichever

1. though, although

though, although 当虽然讲，都不能和 but 连用。Although (though)...but 的格式是不对的，但是他们都可以同 yet (still) 连用。所以 though (although)...yet(still) 的格式是正确的。

Wrong: Although he is rich but he is not happy.
Right: Although he is rich, yet he is not happy.
虽然他很富有，然而他并不快乐。
Right: Although we have grown up, our parents treat us as children.
Right: Although we have grown up, our parents still treat us as children.
尽管我们已经长大了，可是我们的父母仍把我们看作小孩。

2. as, though 引导的倒装句

as / though 引导的让步从句必须表语或状语提前（形容词、副词、分词、实义动词提前）。

Child as /though he was, he knew what was the right thing to do.
= Though he was a small child, he knew what was the right thing to do.
他虽然是个孩子，但他知道什么是正确的事情。

注意：
①句首名词不能带任何冠词。
②句首是实义动词，其他助动词放在主语后。如果实义动词有宾语和状语，随实义动词一起放在主语之前。

Try hard as he will, he never seems able to do the work satisfactorily.
= Though he tries hard, he never seems...
虽然他很努力，但他的工作总做的不尽人意。

四、比较状语从句

常用引导词：as(同级比较)，than(不同程度的比较)

特殊引导词：the more... the more..., just as..., so..., A is to B what /as X is to Y, no... more than, not A so much as B

1. as...as 和……一样

   Jack is as tall as Bob. 捷克和汤姆一样高。

2. not so(as)...as... 和……不一样

   She is not so(as)outgoing as her sister. 她不如她姐姐外向。

3. more...than 比……更……

   This book is more instructive than that one. 这本书比那本书更有教育意义。

五、方式状语从句

常用引导词：as, as if, as though

特殊引导词：the way

When in Rome, do as the Romans do. 入乡随俗。

She behaved as if she were the boss. 她表现得好像她是老板。

as if, as though 两者的意义和用法相同，引出的状语从句谓语多用虚拟语气，表示与事实相反，有时也用陈述语气，表示所说情况是事实或实现的可能性较大。汉语译常作"仿佛……似的""好像……似的"，例如：

He looks as if (as though) he had been struck by lightning.

他那样子就像被闪电击了似的。（与事实相反，谓语用虚拟语气。）

It looks as if the weather may pick up very soon.

看来天气很快就会好起来。（实现的可能性较大，谓语用陈述语气。）

Exercises

Choose the best answer.

1. _____ there's a will, there's a way.
   A. Where   B. wherever   C. when   D. whenever
2. Read it aloud _____ the class can hear you.
   A. so that   B. if   C. when   D. although
3. _____ you go, don't forget your people.
   A. Whenever   B. However   C. Wherever   D. Whichever
4. They will never succeed _____ hard they try.
   A. because   B. however   C. when   D. since
5. _____ still half drunk, he made his way home.
   A. When   B. Because   C. Though   D. As
6. _____ she was very tired, she went on working.
   A. As   B. Although   C. Even   D. In spite of
7. Busy _____ he was, he tried his best to help you.
   A. as   B. when   C. since   D. for

8. She was _____ tired _____ she could not move an inch.
   A. so, that   B. such, that   C. very, that   D. so, as
9. Although it's raining, _____ are still working in the field.
   A. they   B. but they   C. and they   D. so they
10. Speak to him slowly _____ he may understand you better.
    A. since   B. so that   C. for   D. because
11. I hurried _____ I wouldn't be late for class.
    A. since   B. so that   C. as if   D. unless
12. _____, he is good at drawing.
    A. To be a child   B. A child as he is
    C. As a child   D. Child as he is
13. Bring it nearer _____ I may see it better.
    A. although   B. even though   C. so that   D. since
14. Helen listened carefully _____ she might discover exactly what she needed.
    A. in that   B. in order that   C. in case   D. even though
15. More people will eat out in restaurants _____ they do today.
    A. than   B. when   C. while   D. as
16. _____ hard she tries, she can hardly avoid making mistakes in her homework.
    A. Much   B. However   C. As   D. Although
17. Poor _____ it may be, there is no place like home, _____ you may go.
    A. as, wherever   B. though, whenever
    C. in spite of; when   D. that; wherever
18. The history of nursing _____ the history of man.
    A. as old as   B. is old than
    C. that is as old as   D. is as old as
19. After the new technique was introduced, the factory produced _____ tractors in 1988 as the year before.
    A. as twice many   B. as many twice
    C. twice as many   D. twice many as
20. It was _____ that she couldn't finish it by herself.
    A. so difficult a work   B. such a difficult work
    C. so difficult work   D. such difficult work

CULTURE TIPS

扫一扫了解网络管理员考试要求

Unit Ten

IT Jobs

Section One Warming Up

Nowadays there are a lot of IT-related jobs available across the world, and you can find an ideal job via searching on the Internet. If you are a skillful personal in IT area, your job seeking opportunity is great.

Pair work

1. If you are going to apply for a job in an IT company, can you name any positions? Try to match the following English titles with the corresponding Chinese and tell each other what the job descriptions are for each job.

   A. Web Designer   程序员
   B. Web Editor   数据分析
   C. Website Promotion   软件测试工程师
   D. Network Administrator   网络管理员
   E. Data Analyst   网站编辑
   F. Graphic Designer   软件开发
   G. Programmer   网站设计师
   H. Software Developers   平面设计师
   I. Software Testing Engineer 网站推广

2. What should you prepare for the job application? Name the following items and get ready for

the job interview.

Resume   Certificate
ID card   Diploma

1._____

2._____

3._____

4._____

Unit Ten IT Jobs

Section Two Real World

Alex has read about a job opening for a computer programmer advertised by a computer company in *China Daily* and now he is calling the company to ask about it.

Listen and act the following dialogue.

Asking about a Job Opening

Alex: Hello! This is Alex Wong. I'm calling to ask whether you have a job opening for a computer programmer.

Personnel: Yes, there could be an opening in three weeks.

Alex: What are the **qualifications** for the job?

Personnel: The **applicants** should have at least a college education on computer science, and with some work experience. Furthermore, English would also be an **essential requirement** for this job.

Alex: Is the job **temporary** or **permanent**?

Personnel: It's a permanent full-time job.

Alex: What are the working hours?

Personnel: It's from 9 am to 5 pm, five days a week.

Alex: And what about the **salary**?

Personnel: Well, the starting salary is $2,000 a month, with some basic **benefits** such as health **insurance**, sick leave, paid **vacation** and so on.

Alex: OK. I'm interested in this job. And how can I **apply** for the job?

Personnel: You can find the **application** form in our website and fill it in.

Alex: Thank you so much for the information. Goodbye!

Personnel: Goodbye!

Words & Expressions

personnel [pɜːsəˈnel]	n.	the department responsible for hiring, training, placing employees and for setting policies for personnel management 人事部门
qualification [ˌkwɒlɪfɪˈkeɪʃ(ə)n]	n.	an attribute that must be met or complied with and that fits a person for something 资格
applicant [ˈæplɪk(ə)nt]	n.	a person who requests or seeks something such as assistance or employment or admission 申请人
essential [ɪˈsenʃ(ə)l]	adj.	basic and fundamental 基本的；必要的
requirement [rɪˈkwaɪəm(ə)nt]	n.	anything indispensable 要求；必要条件

temporary ['temp(ə)rərɪ]	adj.	not permanent; not lasting 暂时的；临时的
permanent ['pɜːm(ə)nənt]	adj.	continuing or enduring without marked change in status or condition or place 永久的
salary ['sælərɪ]	n.	wage or income 薪水
benefit ['benɪfɪt]	n.	something that aids or promotes well-being 利益；好处
insurance [ɪn'ʃʊər(ə)ns]	n.	promise of reimbursement in the case of loss 保险；保险费
apply [ə'plaɪ]	vt.	put into service; make work or employ for a particular purpose or for its inherent or natural purpose 申请；应用
application [ˌæplɪ'keɪʃ(ə)n]	n.	the work of applying something 申请
a full-time job		全职工作
health insurance		医疗保险
sick leave		病假
paid vacation		带薪休假
application form		申请表

Find Information

Task I Read the dialogue carefully and then answer the following questions.

1. Where is the job that Alex plans to apply for advertised?

2. What's the job title?

3. What are the qualifications of the job?

4. What's the salary?

5. How can Alex apply for the job?

Task II Read the dialogue carefully and then decide whether the following statements are true (T) or false (F).

(　) 1. The job would be opening for a month.
(　) 2. English would also be viewed as a qualification for the job.
(　) 3. It's a full-time job with 8 working hours.
(　) 4. Health insurance, sick leave and paid vacation are not included in the basic benefits.

(　　) 5. Alex is not interested in the job.

Words Building

Task I **Choose the best answer from the four choices A, B, C and D.**

(　　) 1. If you don't _____ smoking, you will never get better.
   A. give off	B. give up	C. give in	D. give out

(　　) 2. What you said reminded me _____ something I read in the newspaper.
   A. for	B. by	C. from	D. of

(　　) 3. The children _____ many times not to get close to the fire.
   A. were being told	B. have been told	C. had been told	D. told

(　　) 4. Being engaged in the research work, Dr. Yang seldom goes anywhere _____ his office.
   A. in place of	B. except to	C. in addition to	D. instead of

(　　) 5. Ships are _____ than planes that people take them mainly for pleasure.
   A. very much slower	B. so much slower
   C. too much slower	D. much more slower

Task II **Fill in each blank with the proper form of the word given.**

1. He is not _____ for the job because of his poor health condition. (qualification)
2. The _____ difficult we're facing will soon ravel out. (temporarily)
3. Can I ask about the vacation and _____ leave? (sickness)
4. Where can I fill in the _____ form?(apply)
5. This gun is designed for one purpose— it's _____ to kill people. (basic)

Task III **Match the following English terms with the equivalent Chinese.**

A — job responsibility	1. (　　) 入职日期
B — starting salary	2. (　　) 年度会议
C — join-date	3. (　　) 项目报告
D — training supervisor	4. (　　) 工作职责
E — annual meeting	5. (　　) 保险与福利
F — marketing plan	6. (　　) 培训主管
G — project report	7. (　　) 生产总监
H — insurance and welfare	8. (　　) 起点工资
I — production director	9. (　　) 紧急事项
J — urgent issues	10. (　　) 营销计划

Cheer up Your Ears

Task I Listen and write down what you've heard. Then read and recite till you can use them fluently.

1. I'm calling to ask whether you have a job opening for a computer _____ .
2. There could be an _____ in three weeks.
3. What are the _____ for the job?
4. The _____ should have at least a college education on computer science, and with some work experience.
5. Furthermore, English would also be an essential _____ for this job.
6. It's a _____ full-time job.
7. The starting salary is $2,000 a month, with some basic benefits such as health _____ , sick leave, paid vacation and so on.
8. And how can I _____ for the job?
9. You can find the _____ form in our website and fill it in.
10. Thank you so much for the _____ .

Task II Listen and fill in the blanks with the missing words and role-play the conversation with your partner.

A: Come in. Welcome to the __1__ . Sit down, please.
B: Thank you.
A: Are you sure you want to __2__ this job? The job is to write __3__ .
B: Sure! I am the person most __4__ for the job.
A: Ah…well, we are looking for someone who can do a really __5__ job.
B: Please give me a chance! I have the __6__ experience at the ABC Computer Company for two years.
A: What did you do there?
B: I was __7__ writing computer programs.
A: Good. Why did you leave your last job?
B: I was __8__ because it was a temporary assignment.
A: A wage suitable to the position with the opportunity to __9__ .
B: Ok. I'll let you know our final __10__ in two weeks.
A: I'll keep my fingers crossed.
B: So thank you for coming today. See you.
A: See you.

Task III Listen to 5 recorded questions and choose the best answer.

1. A. Don't mention it.   B. This way, please.

C. You're right.   D. No, thanks.
2. A. It's over there.   B. Sorry, he isn't in.
   C. That's very kind of you.   D. I'd love to.
3. A. Yes, I will.   B. You're welcome.
   C. I'm sorry to hear that.   D. Not too bad.
4. A. That's too bad.   B. Thanks a lot.
   C. No problem.   D. Here you are.
5. A. Good idea.   B. See you soon.
   C. Never mind.   D. Hold on, please.

Table Talk

Task I Complete the following dialogue by translating Chinese into English orally.

A: Good morning, Mr. Wong. We have here your application for the position of computer programmer. Tell me some key facts about yourself.

B: Sure. I'm a colleague graduate and I majored in ___1___ （软件工程）.

A: Do you have any ___2___ （工作经验）?

B: Yes. I worked as a computer programmer in IBM for six month as an ___3___ （实习生）.

A: What are your strengths?

B: I am reliable and hardworking and a quick learner.

A: Do you speak any foreign language, such as English?

B: Yes, of course. I am good at ___4___ （口头英语） and my second foreign language is Japanese.

A: That's fine. But you will still need to take our English test tomorrow morning.

B: No problem.

A: What salary do you expect?

B: A wage suitable to the position with the ___5___ （升职机会）.

A: Ok. Our final decision will be available next week. Thank you for coming to this interview.

B: Thank you for considering my application.

Task II Pair-work. Role-play a conversation about the real situation of a job interview with your partners.

Situation

Suppose you are college graduate who is now being interviewed by an HR officer. You are concerned about the basic benefits and ask the officer about it.

Section Three Brighten Your Eyes

Pre-reading questions:
1. What's your future plan after graduation?
2. Do you have any idea about IT jobs?

Jobs with the Brightest Prospects

Job seekers with computer **skills** and training are therefore likely to have **plentiful** opportunities. Let's get familiar with some IT related jobs.

Software Engineers

Software engineers create applications for networks, data system, the Web and increasingly, for mobile **devices** like smart phones or **tablets** (such as Apple's iPad). When Apple **brags** that "there's an app for that", the company means that customers have access to **myriad** applications. It also means that the **folks** who design those applications—for just about any company—are in high demand.

Operations Research **Analysts**

The title can sound a little **mysterious** to the **uninitiated**, but basically these workers solve problems using math, computer modeling, programming and other types of **quantitative analysis**. The methods of operational research were developed for the military but are now used by a wide range of public and private organizations. Operations research analysts will see **excellent** job prospects as organizations look for ways to become more **efficient**.

Systems Analysts

Whether they're designing a computer system from **scratch** or **tweaking** an existing one, systems analysts have to be able to see the big picture. The job requires a thorough understanding of how an organization functions, so that the hardware and software they choose can best serve that organization's need.

Words & Expressions

prospect ['prɒspekt]	n.	the possibility of future success 前途；预期
skill [skɪl]	n.	an ability that has been acquired by training 技能；本领
plentiful ['plentɪfʊl]	adj.	existing in great number or quantity 丰富的；许多的
device [dɪ'vaɪs]	n.	an instrumentality invented for a particular purpose 装置
tablet ['tæblɪt]	n.	a slab of stone or wood suitable for bearing an inscription 碑；写字板

brag [bræg]	vi.	show off 吹牛；自夸
myriad ['mɪrɪəd]	adj.	a large indefinite number 无数的；种种的
folk [fəʊk]	n.	people in general (often used in the plural) 民族；人们
analyst ['æn(ə)lɪst]	n.	someone who is skilled at analyzing data 分析者
mysterious [mɪ'stɪərɪəs]	adj.	of an obscure nature 神秘的；不可思议的
uninitiated [ʌnɪ'nɪʃɪeɪtɪd]	adj.	not initiated; deficient in relevant experience 不知情的；缺少经验的
quantitative ['kwɒntɪˌtətɪv]	adj.	expressible as a quantity or relating to or susceptible of measurement 定量的；数量的
analysis [ə'nælɪsɪs]	n.	an investigation of the component parts of a whole and their relations in making up the whole 分析；分解
excellent ['eksələnt]	adj.	of the highest quality 卓越的；极好的；杰出的
efficient [ɪ'fɪʃ(ə)nt]	adj.	being effective without wasting time or effort or expense 有效率的；生效的
scratch [skrætʃ]	n.	an abraded area where the skin is torn or worn off 擦伤；乱写
tweak [twiːk]	vt.	pinch or squeeze sharply 扭；用力拉
get familiar with...		熟悉……
Software Engineer		软件工程师
have access to...		使用；可以利用
in high demand		需求高
Operations Research Analyst		运筹分析师
a wide range of...		大范围的……
Systems Analysts		系统分析员

Find Information

Task 1 Read the passage carefully and then answer the following questions.

1. What does a Software Engineer do?

2. What kind of methods do Operations Research Analysts use to solve problem?

3. Do Operations Research Analysts have an excellent prospect?

4. What do Systems Analysts do in their jobs.

5. What should a Systems Analyst have as a basic requirement?

Task II Read the passage carefully and then decide whether the following statements are true (T) or false (F).

(　　) 1. Job seekers with computer skills and training are unlikely to have lots of chances.
(　　) 2. Software engineers are in high demand.
(　　) 3. Operations research analysts will see good job prospects as organizations look for ways to become more efficient.
(　　) 4. Systems Analysts should have the ability to see bigger picture.
(　　) 5. Hardware and software that best serve one organization's need are chosen by software engineers.

Words Building

Task I Translate the following phrases into English.

1. 移动设备_____
2. 智能手机_____
3. 计算机建模_____
4. 定量分析_____
5. 运筹学_____

Task II Translate the following sentences into Chinese or English.

1. Job seekers with computer skills and training are likely to have plentiful opportunities.
   _____

2. When Apple brags that "there's an app for that", the company means that customers have access to myriad applications.
   _____

3. The methods of operational research were developed for the military but are now used by a wide range of public and private organizations.
   _____

4. Let's get familiar with some IT related jobs.
   _____

5. 软件工程师为网络、数据系统、网站，更多的是为如智能手机或平板电脑等移动设备设计应用程序。
   _____

6. 随着各组织机构对高效的寻求，运筹分析师将有极佳的工作前景。
   _____

7. 不论系统分析员从零开始设计一个计算机系统还是修改一个已存的系统，他们都必须顾全大局。
   _____

8. 这份工作需要对一组织机构如何运作有透彻的理解，这样才能选择出更好地服务于这一机构的硬件及软件。

_____

Task III Choose the best answer from the four choices A, B, C and D.

( ) 1. He doesn't dare to leave the house _____ he should be recognized.
　　A. in case　　　　B. if　　　　C. provided　　　　D. so that
( ) 2. The world _____ we live is made up of water.
　　A. at which　　　B. in which　　C. on which　　　D. of which
( ) 3. As a matter of fact, they would rather leave for China than _____ in America.
　　A. stay　　　　　B. staying　　　C. stayed　　　　D. to stay
( ) 4. _____ one you choose, I am sure you will enjoy it.
　　A. Whatever　　　B. Which　　　C. That　　　　　D. Whichever
( ) 5. The police are _____ this matter.
　　A. looking on　　B. looking into　C. looking at　　　D. looking out

Task IV Fill in each blank with the proper form of the word given.

1. Confronted with a complicated problem, you should ask a _____ co-worker for help. (skill)
2. The wounded need great _____ of food and drinking water. (quantitative)
3. Yesterday is history, tomorrow is a _____ and today is a gift. (mysterious)
4. We should learn to discern and _____ the essentials of complicated questions. (analysis)
5. To improve the _____ of hybrid element for implementation is suggested. (efficient)

Cheer up Your Ears

Task I Listen and write down what you've heard. Then read and recite till you can use them fluently.

1. Job seekers with computer skills and training are therefore likely to have plentiful _____ .
2. Software engineers create _____ for networks, data system, the Web and increasingly, for mobile _____ like smart phones or tablets.
3. It also means that the folks who _____ those applications— for just about any company— are in high _____ .
4. The title can sound a little mysterious to the _____ , but basically these workers solve problems using math, computer modeling, programming and other types of quantitative _____ .
5. The methods of _____ research were developed for the military but are now used by a wide _____ of public and private organizations.
6. Operations research analysts will see excellent job _____ as organizations look for ways to

become more _____ .

7. Whether they're designing a computer system from _____ or tweaking an existing one, systems analysts have to be able to see the big picture.

8. The job requires a thorough understanding of how an organization _____ , so that the hardware and software they choose can best serve that organization's need.

Task II Listen to 5 short dialogues and choose the best answer.

1. A. Customer and client.
   B. Customer and restaurant server.
   C. Customer and shop assistant.
   D. Customer and bank clerk.
2. A. How to open an account.
   B. How to draw money from the bank.
   C. What the different interest rates are.
   D. About all the different accounts.
3. A. To cash his check.
   B. To write his name on the check.
   C. To show his traveler's check.
   D. To pay for cashing his check.
4. A. Mathematics.　　B. Credits.　　C. Banking.　　D. Computer.
5. A. $ 18.50　　B. $ 19.50　　C. $ 15.50　　D. $ 14.00

Table Talk

Group Work

Talk about the IT related jobs with your partner as much as you know.

_____

_____

Section Four Translation Skills

科技英语的翻译方法与技巧——省略句和长句的翻译

一、省略句的译法

为减少或避免用复合句，缩短句子长度，常采用省略句。

1. 并列复合句中的省略

各分句中的相同成分可以省略。例如：

The first treatment would require a minimum of 48 hours, while the second treatment would require only 26 hours.

第一次处理最少需要 48 小时，而第二次处理只需要 26 小时。
可以省略为：
The first treatment would require a minimum of 48 hours, the second only 26 hours.

2. 状语从句中的省略

在复合句中，当状语从句的主语和主句的主语相同时，从句中的主语和助动词可以省略。例如：

[When (they are)] heated under pressure, they constitutes (together).
当在压力下加热时，各成分会熔合到一起。

3. 定语从句中的省略

由关联词 that 或 which 引导的定语从句，可将关联词和从句中的助动词一并省略。例如：

With tin, copper forms a series of alloys which are known as bronze.
铜与锡形成一系列叫做青铜的合金。
可以省略为：
Alloyed with tin, copper forms a bronze.
向其中添入锡，铜就形成了青铜。

4. 其他省略句

（1）通过改变句子结构，可将较长的句子压缩成较简单的句子。例如：

In the form in which they have been presented, the test results give no useful information.
可以省略为：
Thus presented, the results give no useful information.
这样给出的实验结果不能提供什么有用的信息。

又如：
Normally lead was extruded at room temperature, aluminum either cold or hot, and copper hot.
通常，铅在室温下进行挤压，铝可进行冷挤压或热挤压，铜可进行热挤压。

（2）使用标准的缩写词，也可将长句缩短。例如：

The interface composition between matrix and reinforce can be analyzed by EMPA.
应用电子显微镜可以分析基体与增强物之间的界面成分。（EMPA 是 Electron Microprobe Analysis 的首字词）

5. 常见省略句型

As described above ...	如前所述……
As shown in figure 3...	如图 3 所示……
As indicated in table 2...	如表 2 所指示……
As already discussed...	如前面讨论过的……
As noted later...	如后面所说明的……
If any (anything)...	如果有的话，即使需要也……
If convenient...	如果方便的话……
If necessary...	必要时，如果必要的话……
If possible...	如有可能……
If required...	如果需要……

If not...	如果不……
If so...	如果是这样，果真如此……
When in use...	在使用时，当工作时……
When necessary...	必要时……
When needed...	需要时……
When possible...	有可能的……

二、长句的译法

对科技英语中复杂的长句有下列处理方法。

1. 顺译法

长句叙述层次与汉语相近时，可按英语原文顺序依次译出。例如：

They use their knowledge of the working properties of metals and their skills with machine tools to plan and carry out the operations needed to make machined products that meet precise specifications.

他们运用其金属材料特征方面的知识和机床方面的技能进行工艺规划和加工，制造满足精度要求的机加工产品。

Since this form of CNC machine can perform multiple operations in a single program (as many as CNC machines can), the beginner should also know the basics of how to process workpieces machined by turning so a sequence of machining operations can be developed for workpieces to be machined.

由于这种数控机床能在一个程序中完成多种加工（很多数控机床都能够如此），初学者同时应该了解如何加工车削类零件的基础知识，这样才能编制出待加工零件的加工工序。

2. 逆译法

有时英语长句的展开层次与汉语表达方式相反，这时就需要逆着原文的顺序译出。英语的表达习惯是先说出主要的，然后才说次要的。

Therefore, the number of workers learning to be machinists is expected to be less than the number of job openings arising each year from the need to replace experienced machinists who retire or transfer to other occupations.

每年都有一些有经验的机械师退休或跳槽到其他职业，因此带来的工作空缺数量大于准备从事机械工工作的工人数量。

HAAS advises that you not change parameters unless you know exactly what needs to be changed and why, and that you have made all the correct inquiries within your shop and with HAAS service personnel.

HAAS 公司建议用户，在无法确定需要改什么和为什么要改，以及没有正确咨询车间和 HAAS 服务人员之前，不要随意更改参数。

3. 分译法

有时英语长句中嵌套多重定语从句，或者英语长句中各个主要概念在意义上并无密切联系时，可以拆成独立的短句，再按照汉语习惯重新安排次序。

The headstock is required to be made as robust as possible due to the cutting forces involved, which can distort a lightly built housing, and induce harmonic vibrations that will transfer through to the workpiece, reducing the quality of the finished workpiece.

床头箱的制造必须十分坚固，因为切削力的影响可能会使制造不坚固的机座变形，并且产生谐振，谐振传至工件，会降低成品件的质量。

Listening

Task I Listen to 2 conversations and choose the best answer.

Conversation 1

1. A. At a supermarket.　　　　B. At a bank.
   C. At a bookstore.　　　　　D. At a travel agency.
2. A. Fifty dollars.　　　　　　B. Ten dollars.
   C. Twenty dollars.　　　　　D. Twelve dollars.

Conversation 2

3. A. A dark green sweater.　　B. A black coat.
   C. A white shirt.　　　　　　D. A green jacket.
4. A. A6587T.　　　　　　　　　B. A7678T.
   C. A4578T.　　　　　　　　　D. A5687T.
5. A. With an American Express card.
   B. With a traveler's check.
   C. With a Master card.
   D. With a VISA card.

Task II Listen to the passage and fill in the blanks with the missing words or phrases.

Many cultures have different ideas about why people catch colds. For example, in the United States, some people think that you can catch cold if your ___1___ get cold. So, mothers tell small children to wear ___2___ boots in the winter. In other places, including parts of the Middle East, some people believe that strong winds cause colds. So, on trains and buses, people usually don't like to sit ___3___ open windows. In parts of Europe, some people think that wearing wet clothes will give you a cold. They say that after you go ___4___ , you should quickly put on dry clothes. Today, scientists know that colds are caused by viruses. But the cold ideas are still very strong, and many people still follow them to ___5___ getting ill.

Extensive Reading

Directions: Read the passage carefully and then decide whether the following statements are true (T) or false (F).

Ten Interview Questions

It's important to be prepared to respond effectively to the interview questions that employers typically ask at job interviews. Since these questions are so common, hiring managers will expect you to be able to answer them smoothly and without hesitation. Review the top 10 interview questions you'll most likely be asked at a job interview, plus the best answers.

1. What is your greatest strength?

When you are asked about your greatest strengths, it's important to discuss the attributes that will qualify you for the specific job and set you apart from the other candidates.

2. What is your greatest weakness?

Do your best to frame your answers around positive aspects of your skills and abilities as an employee.

3. Why are you leaving or have left your job?

Stick with the facts, be direct and focus your interview answer on the future, especially if your leaving wasn't under the best of circumstances.

4. Tell me about yourself.

Start by sharing some of your personal interests which don't relate directly to work.

5. Why do you want this job?

Be specific about what makes you a good fit for this role, and mention aspects of the company and position that appeal to you.

6. Why should we hire you?

Make your response a concise sales pitch that explains what you have to offer the employer, and why you should get the job.

7. How do you handle stress and pressure?

The best way to respond to this question is to give an example of how you have handled stress in a previous job.

8. Describe a difficult work situation/project and how you overcame it.

As with the question about stress, be prepared to share an example of what you did in a tough situation.

9. How do you evaluate success?

A question like this gives your potential employer a sense of your work ethic, your goals, and your overall personality. Consider the company and your role and formulate an answer based on those and your personal values and goals.

10. What are your goals for the future?

Keep your answer focused on the job and the company you're interviewing with.

(　　) 1. The interview questions are not so common that hiring managers will expect you to be able to answer them smoothly.

(　　) 2. When you are asked about your greatest strengths, you should discuss the attributes set you apart from the other candidates.

(　　) 3. When you are asked about your weakness, do your best to frame your answers around positive aspects.

(　　) 4. When talk about yourself, share your personal interests which relate directly to work.

(　　) 5. Stick to the fact and be direct when asked about your leaving.

(　　) 6. To illustrate how you handle stress, you'd better give an example of how you have handled stress in a current job.

(　　) 7. "Describe a difficult work situation and how you overcame it." is a question about stress.

(　　) 8. Goals for the future should be focused on the job and the company you're interviewing with.

Section Five Writing

Agreement and Contract 协议与合同

协议与合同是指在共同办理某事中，为了明确当事人双方或数方在这一事务中的责任、义务和权利，经相互讨论协商，双方或所涉及方所签署的具有法律效应的书面正式文件。一般来说，在涉及教育合作、技术服务、物流运输等方面的文件时，agreement 用得比较多；在涉及贸易往来、租赁、劳工关系等方面的文件时，contract 一词用得较为普遍。

Sample

Cooperation Agreement

Agreement No. JM008　　　　　　　　　　　　　　Date: August 6, 2007
Signed At: Beijing
Party A: Beijing Xiehe Hospital
Party B: Singapore National University Hospital

To further promote the cooperation between the two parties, Party A and Party B have on the 6th of August, 2007 reached the following understanding:

1. Party A will provide Party B with the practice chances for experiment training. Party B will provide Party A with English language training programs of various levels.

2. The two parties will exchange their doctors and visiting scholars, based on the necessity of academic exchange.

3. Party A and Party B will cooperate to establish a research institute of pharmacology. They will also develop other programs which will mutually benefit the social, economic and educational development of the two respective countries.

This memorandum of understanding is written in both English and Chinese.

Party A Representative:　　　　　　Party B Representative:
Signature:_____　　　　　　　Signature:_____

合作协议

协议编号：JM008
签约地点：北京
甲方：北京协和医院
乙方：新加坡国大医院

> 为促进双方的进一步合作，甲方与乙方于 2007 年 8 月 6 日达成以下共识：
> 1. 甲方给乙方提供试验实践培训机会，乙方给甲方提供不同水平的英语培训。
> 2. 根据学术交流的需要，两方互派医生和访问学者。
> 3. 双方合办药学专业研究所，发展双方所在国均可受益的社会、经济、教育等项目。
> 本协议备忘录以中英两种文字写成。
>
> 甲方代表：（签字）　　　　　　　　　　　　　　乙方代表：（签字）
>
> 　　　　　　　　　　　　　　　　　　　　　　　　2007 年 8 月 6 日

写作格式：

协议或合同有多种不同的形式，可以根据实际情况灵活变通。协议结构较为简单，而一份正式的贸易合同结构十分严谨。协议或合同通常由开头部分、正文和结尾组成。

开头部分：包括合同名称、合同编号、签约时间和地点、签约双方的全称及签约目的等。

正文：正文是合同的主要部分，包含了协议或合同的内容。正文明确规定当事人的具体权力和义务、违约赔偿、争议解决、适用法律等条款。

结尾：通常包括文字效力、份数、当事人签字、盖印等。

协议或合同的语言应正式规范，行文严谨，措词准确以避免误解和歧义。

常用句型：

1. This Contract is signed in ＿＿＿（place）on ＿＿＿（day, month, year）, by and between ＿＿＿ on one hand and ＿＿＿ on the other hand.

本合同由 ＿＿＿ 为一方和 ＿＿＿ 为另一方于 ＿＿＿（时间）在 ＿＿＿（地点）签订。

2. In accordance with ＿＿＿, adhering to ＿＿＿ and through ＿＿＿, both parties agree ＿＿＿. The Contract is worked out thereunder.

根据 ＿＿＿，本着 ＿＿＿，经过 ＿＿＿，双方同意 ＿＿＿。兹订立本合同。

3. The Buyers agree to buy and the Sellers agree to sell the following goods on terms and conditions set forth below.

买卖双方同意按下列条款由买方购进卖方售出下列商品。

4. All disputes arising out of this Contract shall be referred to the International Court of Law.

涉及本合同的一切争议将提交国际法庭裁决。

5. This Contract is made out in duplicate in Chinese and English, one Chinese original copy and one English original copy for each Party, both texts being equally authentic.

本合同用中文和英文书就，一式两份，双方各执一份中文正本和一份英文正本。两种文本具有同等效力。

6. This Contract shall come into force from the date it is signed by authorized representatives of both parties.

本合同自双方授权代表签字之日起生效。

7. The two parties shall settle disputes through friendly negotiations.

如有争议，双方应协商解决。

8. The two parties shall strictly follow the items in this agreement.

双方均应严格执行本合同所有条款。

9. If the foreign teacher goes without permission this school has the option to consider the contract violated if they wish.

如外教未经许可擅离职守，学校有权以违约视之。

10. Either party upon 60-day written notice may terminate this agreement.

任何一方可提前60天以书面通知终止协议。

Practice

Complete the following sentences according to the given information in Chinese. 根据中文提示和上下文内容填写相应的词语。

1. The foreign teacher shall _____ the following _____ . This school provides these benefits _____ the foreign teacher. No _____ given to those who do not choose to participate or avail themselves of these benefits.

外教可享受以下其他待遇。校方向外教无偿提供这些待遇。如外教自动放弃这些待遇，校方无须提供任何补偿。

2. The foreign teacher shall _____ if unable to work _____ the concerned class time. This school may require the foreign teacher to see a doctor prior to _____ . Teachers will be allowed _____ . The foreign teacher may not _____ unused sick days.

外教如因生病无法工作应在相关上课时间之前提交书面病假申请。校方可在准假前要求外教去就诊。校方允许外教每月有2天带薪病假时间。未使用的带薪病假时间不得积累。

Section Six Grammar

Substantive clause 名词性从句

一、名词性从句的基本概念

在句子中起名词作用的句子叫名词性从句 (Substantive clause)。名词性从句的功能相当于名词词组，它在复合句中能担任主语、宾语、表语、同位语、介词宾语等，因此根据它在句中不同的语法功能，名词性从句又可分别称为主语从句、宾语从句、表语从句和同位语从句。

二、引导名词性从句的连接词

引导名词性从句的连接词可分为以下3类。

1. 从属连词

that（无任何词意）；whether, if（均表示"是否"表明从句内容的不确定性）；as if, as though（均表示"好像""似乎"）。

以上连词在从句中均不充当任何成分。

2. 连接代词

what, whatever, who, whoever, whom, whose, which, whichever, whomever

3. 连接副词

when, where, how, why

大部分连接词引导的主语从句都可以置于句末，用 it 充当形式主语。

It is not important who will go. 谁去并不重要。

It is still unknown which team will win the match.

哪个队会赢得比赛还不得而知。

三、具体分类

1. 主语从句

作句子主语的从句叫主语从句。主语从句通常由从属连词 that，whether，if 和连接代词 what，who，which，whatever，whoever 以及连接副词 how，when，where，why 等词引导。that 在句中无词义，只起连接作用；连接代词和连接副词在句中既保留自己的疑问含义、又起连接作用，在从句中充当从句的成分。

例如：

What he wants to tell us is not clear. 他要跟我们说什么，还不清楚。

Who will win the match is still unknown. 谁能赢得这场比赛还不得而知。

It is known to us how he became a writer.

我们都知道他是如何成为一名作家的。

Where the English evening will be held has not yet been announced.

英语晚会将在哪里举行，还没有宣布。

2. 宾语从句

用作宾语的名词性从句叫宾语从句。引导宾语从句的关联词与引导主语从句表语从句的关联词大致一样，在句中可以作谓语动词或介词及非谓语动词的宾语。

（1）由连接词 that 引导的宾语从句

由连接词 that 引导宾语从句时，that 在句中不担任任何成分，在口语或非正式的文体中常被省去，但如从句是并列句时，第二个分句前的 that 不可省。

例如：

He has told me that he will go to Shanghai tomorrow.

他已经告诉我他明天要去上海。

We must never think (that) we are good in everything while others are good in nothing.

我们决不能认为自己什么都好，别人什么都不好。

注意：在 demand, order, suggest, decide, insist, desire, request, command 等表示要求、命令、建议、决定等意义的动词后，宾语从句常用"（should）+ 动词原形"。例如：

I insist that she (should) do her work alone. 我坚持要她自己工作。

The commander ordered that troops (should) set off at once.

司令员命令部队马上出发。

（2）用 who，whom, which, whose, what, when, where, why, how, whoever, whatever, whichever 等关联词引导的宾语从句相当于特殊疑问句，应注意句子语序要用陈述语序。

例如：

I want to know what he has told you. 我想知道他告诉了你什么。

She always thinks of how she can work well.

她总是在想怎样能把工作做好。
She will give whoever needs help a warm support.
凡需要帮助的人，她都会给予热情的支持。

3. 表语从句

在句中作表语的从句叫表语从句。引导表语从句的关联词与引导主语从句的关联词大致一样，表语从句位于系动词后，有时用 as if 引导。其基本结构为：主语＋系动词＋that 从句。例如：

The fact is that we have lost the game. 事实是我们已经输了这场比赛。

That's just what I want. 这正是我想要的。

This is where our problem lies. 这就是我们的问题所在。

It looks as if it is going to rain. 看起来要下雨了。

4. 同位语从句

同位语从句说明其前面的名词的具体内容。同位语从句通常由 that 引导，可用于同位语从句的名词有 advice，demand，doubt，fact，hope，idea，information，message，news，order，problem，promise，question，request，suggestion，truth，wish，word 等。

例如：

The news that we won the game is exciting.

我们赢得这场比赛的消息令人激动。

I have no idea when he will come back home. 我不知道他什么时候回来。

The thought came to him that Mary had probably fallen ill.

他想到可能玛丽生病了。

Exercises

Choose the best answer to each of the following items.

1. A warm thought suddenly came to me _____ I might use the pocket money to buy some flowers for my mother's birthday.
   A. if   B. when   C. that   D. which

2. See the flag on top of the building? That was _____ we did this morning.
   A when   B. which   C. where   D. what

3. The government has announced that a modern city will be set up in _____ is still a wasteland now.
   A. what   B. which   C. that   D. where

4. Many people wrote articles on _____ Liu Xiang had failed to compete in the event.
   A. why   B. what   C. who   D. that

5. The couple are spending their holiday on _____ is described as one of the most beautiful islands.
   A. that   B. what   C. which   D. where

6. The book is meant to _____ needs it.
   A. who   B. whoever   C. whomever   D. whom

7. In his speech, Premier Wen Jiabao points out that creativity is _____ it takes to keep a nation highly competitive.
   A. how   B. what   C. which   D. that
8. The experience of the Chinese nation attests to a truth _____ a nation loses in times of disaster will be made up for by her progress.
   A. that what   B. what   C. that   D. what that
9. _____ has recently been done to provide more buses for the people, a shortage of public vehicles remains a serious problem.
   A. That   B. What   C. In spite of what D. Though what
10. _____ is certain is _____ prevention is more important than treatment.
    A. It; that   B. What; that   C. As; what   D. What; what
11. Nobody would stand out admitting the fact, for some reason, _____ they lost the game.
    A. that   B. which   C. what   D. why
12. —The patient looks much better. _____ is it that has made him _____ he is today?
    —Perhaps the special medicine and his family's patient care.
    A. What; that   B. That; that   C. What; what   D. What; which
13. After three hours' climbing, they reached _____ they thought was the place they'd been dreaming of.
    A. what   B. which   C. where   D. that
14. A plan has been put forward _____ more graduates should go to work in the country.
    A. when   B. that   C. whether   D. how
15. It is pretty well understood _____ controls the flow of carbon dioxide in and out the atmosphere today.
    A. that   B. when   C. what   D. how
16. She is very dear to us. We have been prepared to do _____ it takes to save her life.
    A. whichever   B. however   C. whatever   D. whoever
17. The how-to book can be of help to _____ wants to do the job.
    A. who   B. whomever   C. no matter who   D. whoever
18. A good friend of mine from _____ I was born showed up at my home right before I left for Beijing.
    A. how   B. whom   C. when   D. which
19. Many young people in the West are expected to leave _____ could be life's most important decision-marriage-almost entirely up to luck.
    A. as   B. that   C. which   D. what
20. Is there any possibility _____ you could pick me up at the airport?
    —No problem.
    A. when   B. that   C. whether   D. what

CULTURE TIPS

扫一扫了解面试程序

参考文献

1. 陆效用. 通用职场英语 [M]. 大连：大连理工大学出版社，2013.
2. 王蕾，孟建敏. 计算机英语 [M]. 西安：西北工业大学出版社，2005.
3. 谭新星；段琢华. IT 行业英语 [M]. 广州：暨南大学出版社，2012.
4. 张海波. 计算机英语 [M]. 北京：外语教学与研究出版社，2007.
5. 熊英. 实用英语 [M]. 大连：大连理工大学出版社，2009.
6. 刘子轶. 计算机专业英语 [M].2 版. 北京：中国劳动社会保障出版社，2012.
7. 于洪福. 实用英语 [M]. 北京：机械工业出版社，2008.
8. 谈芳，吴云. 高等学校英语应用文写作 [M]. 上海：学林出版社，2005.
9. 惠宇. 新世纪汉英大词典 [M]. 北京：外语教学与研究出版社，2004.
10. 钱书能. 英汉翻译技巧 [M]. 北京：对外经贸大学出版社，2012.